Praise for *The Faith Equation*

This is a book of the kind of **wacky genius one is** .... isions during his or her lifetime.

Peck
veled

Professor Marvin Bittinger uses his powerful kn............................... new **understanding, fresh appreciation and unique pe**................................. n the **Holy Bible.** Where so many other teachers have tak.... ....ders down well beaten paths toward arriving at greater Biblical understanding, Professor Bittinger strikes out on his own through mathematics and delivers a slight turn in the kaleidoscope that **produces an even more brilliant picture of what the Bible truly means.**

Stephen R. Bolt, President of Faith Financial Services, Inc.
Author of *Money On Loan From God*

Marvin L. Bittinger is—likely **the most successful higher-education mathematics author of all time.** Marv's three-book series on developmental mathematics created something entirely new in mathematics education. Most mathematics authors improve on an existing subject or method, but Marv actually created an entirely new approach to remedial college teaching with his series of paperback books.

Marv's textbooks are renowned for their clarity, accuracy, and organization... Most importantly, Marv's texts have enabled millions of students, who may have struggled to learn mathematics previously, to succeed in mathematics.

Here the reader sees another aspect of Marv, his interest in connecting mathematics to the Christian faith.

Greg Tobin, VP/Publisher, Editorial Director
Addison-Wesley Publishing Co.

You are about to undertake **a fascinating journey**—your guide is a man who entered the world of mathematics with little confidence, only to become one of the leading educators of our time, a teacher who developed exciting new methods to help millions of young students find success. Your journey will be similar to those students—and equally exciting—as Marvin Bittinger relates his professional expertise to his religious convictions. Please join him in this wonderful journey.

Charles D. Taylor
Author of *Boomer* and *First Salvo*

Marvin L. Bittinger's *The Faith Equation* is **a remarkably creative and innovative work of wonder and whimsy, precision and paradox, meaning and mystery.** This is an extraordinary effort that combines Dr. Bittinger's successful career in mathematics, his fundamental life message and skills, and his commitment to the mystery we call the triune God of the Bible. *The Faith Equation* adds a number of important insights and twists in its exploration of the mathematical implications of the Christian worldview, and it is a welcome addition to the literature of Christian apologetics.

Kenneth Boa
President, Reflections Ministries, Atlanta, GA
President, Trinity House Publishers, Atlanta, GA
Co-author of *20 Compelling Evidences that God Exists*

The Faith Equation is **a delightful reflection on the spiritual journey and mathematical insights that enriched this journey** by one of the greatest authors of college level mathematics textbooks of our time. The transparency of the author in sharing his own challenges to deal with the paradoxes one encounters in life and in the Christian faith was very refreshing. The use of mathematical ideas such as higher dimensions provides some interesting, if speculative ways of thinking about some of the extraordinary claims one encounters in the Bible. **It is a very enjoyable synthesis of two topics which are seldom addressed, mathematics and Christian faith.**

Dr. Walter Bradley
Professor of Mechanical Engineering, Baylor University

Co-author of *The Mystery of Life's Origin*

THE FAITH EQUATION

THE FAITH EQUATION:

One Mathematician's Journey in Christianity

Marvin L. Bittinger, Ph.D.

LITERARY
ARCHITECTS

International Standard Book Number: 1-933669-07-6

ISBN-13: 978010933669-07-6

Library of Congress Catalog Card Number: Available upon request.

Printed in the United States of America. This book is printed on acid-free paper.

For information on publicity, marketing, or selling this book, contact Bryan Gambrel, Marketing Director, Literary Architects, 317-462-6329.

www.literaryarchitects.com
www.thefaithequation.com

Cover design: GrafikNature
Interior design: Amy Hassos-Parker

To my precious, wife Elaine. You said "yes" that night and captured my heart. And now, after these many years, you have stood beside me as I captured God's dream for me to write this book.

CONTENTS

Acknowledgments

This book began with the vision of Greg Tobin, Vice President, Addison-Wesley Publishing Company. Originally, this writing was part of a companion title, *One Man's Journey Through Mathematics*, published by AW in 2004. But it grew into a separate entity with a life of its own, and so I decided to separate the writing in creation of this book now published by Literary Architects, LLC, of Indianapolis, IN. Special thanks to Renee Wilmeth and Bryan Gambrel for their faith and creative assistance in this, my most favorite publication.

From my favorite math professor of all-time, John K. Baumgart, I first heard the notion that a belief in the existence of God is a *faith axiom*. My fascination with the idea of integrating mathematics and theology thereby started with what might seem like a simplistic bit of information that had a profound impact on my thinking. My brain, my math, and my theology could somehow come together. In hindsight, God was sowing the first seed of this book.

Charles D. "Chuck" Taylor played the part of my developmental editor and protagonist. Chuck brought the wisdom of his years as an editor, a publisher, and an author to bear in advising me throughout the book's writing. To enable me to fulfill this dream, Chuck sacrificed time he could have used to write another novel. Most of all, Chuck was an encourager. You would think I would have more confidence about my writing after 195+ textbooks, but Chuck continually prodded me by saying, "I want more Marv Bittinger in here!" Often, my reaction was, "I don't have anything to say!" Then, at Chuck's urging, and with the grace of God's loving creativity, out it came.

Thanks to M. Scott Peck, who granted me his time and wisdom in three interviews. I thank him for the insight of his years as an author, a psychiatrist, a philosopher, and a Christian. His perception of the notion of paradox is the cornerstone of this book. I'm deeply saddened that his unfortunate death prevents him holding the finished product in his hands.

Special thanks as well, to Dr. Randolph C. Byrd, Dr. William C. Harris, and Dr. David R. Hodge for interrupting their time to grace mine with extensive interviews on the studies discussed in Chapter 6 on the power of prayer. That chapter would not have been as thorough without their special help.

Two outstanding statisticians from the math department at Indiana University-Purdue University Indianapolis (IUPUI) contributed their expertise to this book. In Chapter 4, Dr. Robert Kleyle validated the Time Principle and advised me regarding the issue of independent events. The research of Dr. Jyoti Sarkar on the growth models in Chapter 5 brought in-depth statistical insight to the curve fitting.

Dr. James S. Biddle read many drafts of the manuscript and was always there for me for discussion and encouragement. Just as when I was in graduate school, his special counsel was invaluable.

Ann Fleming, reference librarian at IUPUI, provided prompt, thorough research on many topics. I can't thank her enough for her special assistance.

To all my reviewers, thank you for the pertinent criticisms provided against the background of being close friends. The mathematicians were Dr. Matthew Hassett, Dr. Edward Zeidman, Dr. David Neuhouser, Dr. David Hodge, Barbara Johnson, Judy Beecher, Al Archambault, and Sybil MacBeth. Many medical doctors provided willing counsel. They included Dr. Steven R. Smith, Dr. Bruce F. Schilt, and Dr. John H. Isch.

Others provided extensive review of the theological aspects of the book. They were Dr. John H. Gingrich, Dr. James Livengood, George Donelson, and my pastor, Dave Rodriguez.

To all my encouragers, I appreciate the kind words of support and interest and simply your listening. The list begins with the servanthood of my dear wife Elaine S. Bittinger. She helped with the Biblical research, she came up with ideas, she ran errands to the library, she took phone calls, and, most important, she kept me upbeat and persevering.

Other encouragers were my sons Lowell D. Bittinger and Christopher N. Bittinger, together with their wives Karen E. Bittinger and Tricia A. Bittinger. To these I add Steve Znachko, Jeff Daratony, Glenn Just, Rick Hove, Jim Rathbun, Tom Redmond, Stephen Bolt, Kenneth Boa, Walter Bradley, Benzion Boukai, Todd Pastor, Russ and Susan Daniels, Steve and Jody LaMotte, Danny and Darlene Klingensmith,

Carole and Larry Mattler, Richard and Lorraine Bonewitz, and all my spiritual compatriots in our Wednesday morning theology book group (The Grace-Men): Joe Graves, Mark Fugate, Steve Hockett, Mark Rexroth, Jay Chambers, and Joe Williamson.

I am indebted to Michael J. Rosenborg for the linguistics research in Chapter 10. With a great sense of pride, I must relate that Mike found one of my Algebra-Trig books in a college bookstore, began studying mathematics on his own, went on to receive a Master's Degree in math, and is presently teaching math at a Christian High School in Canyonville, Oregon.

Finally, there is that entity so mysterious, so infinite, so empowering, so demanding, so impossible to describe, and so anticipated. I thank you, God, for the passion of your empowerment! I pray that it fulfills the dream you placed in my heart. You kept me seeking, probing, and asking questions, enough so that I could write this book. I can't wait to sit down with you and get *all* those questions answered.

Introduction: The Beginning of a Mathematician's Journey

In his most famous book, *The Screwtape Letters*, C. S. Lewis writes about Screwtape, an older servant of the deceitful devil, training his young and aspiring demon nephew, Wormwood:

> Above all, do not attempt to use science [mathematics, in our case] as a defense against Christianity. This will positively encourage him to think about realities he can't touch and see.... If he must dabble in science, keep him on economics and sociology.... But the best of all is to let him read no science [and no mathematics] but to give him a grand general idea that he knows it all and that everything he happens to have picked up in his casual talk and reading is "the results of modern investigations."

Before you begin, there are a few things I want to share about my background, my own personal journey, and my own arrival at the faith equation. Consider it background, if you will, but my own journey has been critically important in the creation of this book.

The Faith Equation: One Mathematician's Journey in Christianity can be thought of as using mathematics to enhance a person's reasoning to relate to God. This book is a companion to my memoir, *One Man's Journey Through Mathematic* (Addison Wesley, 2004), an account of my personal experiences over 38 years as an author of mathematics textbooks for college students. This book is dedicated to the premise that mathematicians can integrate mathematics and theology into their lives and professions. But in expanding my ideas about using mathematics as a defense of Christianity, we raise the first question. What is the faith equation?

The Faith Equation

"What is the faith equation?" Skeptics and those with a math background will be quick to ask this question. I know I would look for the equation the first time I opened the book. A person who becomes a follower of Christ inevitably comes to a faith decision of the *mind*, the *heart*, and the *will*. Thus, I define the faith equation to be

Faith = (Mind) + (Heart) + (Will)

The faith equation is more allegorical, like a memory device, rather than an equation involving polynomial, exponential, or trigonometric expressions, although these tools will be used throughout the book. My *mind* is engaged with the equation as it gathers evidence for the decision. My *heart* is engaged in the equation when, despite intellectual hurdles, I feel the need to fill a void in my life with God or by faith. My *heart* is involved when I know within me that there is a disconnect for which life is not providing an answer. In short, my heart tells me I hurt, I feel, I need, I long.... The *will* is engaged as I choose to come to faith—it is a choice that seems to be unequivocal, accompanied by a commitment and a bias to action. In this book, the evidence for the mind is presented primarily in a framework of mathematics, although admittedly an immense amount of evidence lies outside the realm of mathematics.

Numerous theologians, among them Dallas Willard, J. P. Moreland, John Stott, and William Lane Craig, implore Christians to admit the mind into their faith. Moreland writes, "... how unusual it is for Christian people to be taught how to think carefully and deeply about what they believe and why they believe it.... Judged by the scriptures, church history, and common sense, it is clear that something has gone desperately wrong with our modern understanding of the value of reason and intellectual development for individual discipleship and corporate church life." For non-Christians, Moreland also asserts, "... faith is built on reason. We [non-Christians] should have good reasons for thinking that Christianity is true before we [non-Christians] dedicate ourselves completely to it."

I so support the views of these theologians that I have coined the following as a motto for this book:

> God says, "I gave you a brain! Use it in my defense! I know I can stand the test!"

The Unfortunate Divorce of Science and Religion

Those looking for an extensive and critical interpretation of the Bible and its historic veracity will not find it here, although there are wide-ranging references to Biblical topics. Those looking for an in-depth treatment of the philosophical foundations of mathematics will not find it here, nor will they find an axiomatic mathematical system that seeks to prove theorems regarding Christianity. What you will find is one mathematician's journey as he strengthens his faith in a framework of mathematics. The journey will often take a story approach as I intersperse day-to-day spiritual struggles amid situations where mathematics can be brought to bear in defense of the faith. Part of my life as an author of mathematics textbooks is much like a cat on the prowl looking for applications of mathematics. It was inevitable that mathematics would find its way into my studies of the Bible, theology, and apologetics (defense of the faith).

Millions of people study mathematics in high school and college. The least amount of mathematics needed to encounter this book is probably a course in college algebra. Even those with less math than this can grasp the effect of a large exponent in a denominator or the intersection of two graphs, or imagine the idea of a higher dimension. In short, it is possible to gather lots of meaningful Christian apologetics in this book by "reading around the math." In spite of this, there is enough math here to satisfy or entice those blessed with skills of higher mathematics.

The Greek philosophers Plato and Aristotle were credited with the *marriage*, or integration, of science and religion. Science showed them that the heavens and the earth certified the existence of God. The solar system was like a finite, self-contained structure in which God seemed understandable. But the marriage began to crumble in 1572 when Tycho Brahe discovered a new star (actually a supernova) in the constellation Cassiopeia; then, five years later, he observed a new comet in the heavens. A *supernova,* one of the most energetic explosive events known in astronomy, occurs at the end of a star's lifetime, when its nuclear fuel is exhausted and it is no longer supported by the release of nuclear energy. The middle "star" in the belt of the constellation Orion is a supernova.

Subsequent discoveries by Kepler that the orbits of planets were elliptical, by Copernicus and Galileo that the planets revolve around the Sun, and others of Brahe crumbled the marriage of science and religion even further. Paradoxically, the marriage was dissolved by Sir Isaac Newton, 1642-1727, a man as occupied with his Christianity as with science and mathematics, when he developed his Universal Law of Gravity which established that all particles of matter in the universe feel a force of attraction between each other, not just from the Earth.

The growth of scientific knowledge flourished; but in the process, humans—carried away with newfound intellectual power—began to conjure the notion that they could figure it all out by themselves and no longer needed a concept of God. In effect, science became a god unto itself. Instead of pursuing God, man pursued science; science became its own false idol, a false infinite, so to speak. Dallas Willard in his book *Hearing God* asserts in frustration,

---

Today it is simply assumed that scientific knowledge excludes the presence of God from the material universe of which we human beings are supposed to be a pitifully small and insignificant part. This is called *naturalism.* The discoveries of the immensity of space and of the forces of nature—which appear to determine everything that happens and *seem* to run their course with no assistance from the hand of a personal God—can be quite overwhelming.

When the great French mathematician and astronomer Pierre Simon de Laplace [1749-1827] presented Emperor Napoleon with a copy of his book on celestial mechanics, the emperor asked him where God, a supernatural being beyond the natural world, fit into his system. Laplace indignantly drew himself up and replied, "Sir, I have no need of any such hypothesis!" According to the current model of natural sciences, nature proceeds without invoking God.... [T]he university system stands in world culture as the source of unquestioned authority so far as *knowledge* is concerned.... [I]t currently throws its weight behind a picture of reality without God, a picture in which human beings are entirely on their own. Regardless of what the recognized system of education might say of itself for public relations purposes, it presumes in its processes that you can have the best education possible and be ignorant of God.

Denis Alexander writes in his recent book *Rebuilding the Matrix*, "the scientific enterprise is full of experts on specialist areas but woefully short of people with a unified worldview."

Deism is the belief held by a person who does not embrace any particular religious faith but does believe in God, a God who for the most part created the world and left it to its own devices. *Theism*, according to Alexander, is "the view held by the Keplers, Galileos, and Boyles of the 17th century that God not only created the universe in the beginning, but is in a moment by moment relationship with it, actively upholding and sustaining both its existence and its properties. In 'theism' the universe is viewed as contingent ('dependent') upon God's continuing creative actions, whereas in 'deism' the universe is noncontingent ('independent')."

Alexander states further that, "Science and mathematics can go beyond religious, cultural, and political differences to explore the universe and all its workings, but it cannot attach meaning and purpose to such knowledge. But, the faith of a theistic scientist, or mathematician, can attach meaning and purpose which transcends this knowledge.... [T]heism provides a unified worldview that does a remarkably effective job in providing a matrix for science in which the validity of scientific knowledge is justified and in which the fruits of scientific discoveries are channeled in ways that affirm human value, justice and care for the environment."

My passion and profession is mathematics, and my faith is Christian. It is my goal in this book to integrate theology and mathematics as evidence of the Christian faith. *A topic is considered only if it can be presented in a framework of mathematics.* Because mathematics is the queen or handmaiden of the sciences, there will be inevitable references and examples from science, but in this book mathematics will be in the forefront.

Thus, the audience for this book includes anyone who seeks to explore the Christian faith and is curious enough to enjoy my expansion of it in a mathematical framework. Excellent books by Stephen Hawking, Brian Greene, Michio Kaku, and Hugh Ross manage to bring what seem like advanced topics such as quantum physics, string theory, and relativity within the grasp of the ordinary reader. My goal in this book is to bring topics like college algebra, probability, statistics, and a slight amount of calculus within the grasp of the ordinary reader in an examination of Christian theology. In situations where the math level may seem high, I encourage you to read around the mathematics to get the ideas and results.

Using mathematics as a framework for spiritual insight, the following topics will be considered:

- Faith axioms
- Paradox and psycho-spiritual growth
- Probability and prophecy
- Modeling the growth of Christian evangelism
- The power of prayer
- Higher dimensions
- Numerical applications of mathematics in the Bible
- Gödel's Incompleteness Theorem as a metaphor in faith

The topic of paradox is more philosophical, because it integrates mathematical logic and theology to create a cornerstone for the rest of the book.

The Story Behind this Book

I have found joy in the belief that God blessed each of us with a wonderful brain and fully expects us to use it to understand and enhance the Christian faith. It is common religious jargon to say, "God told me that!" Although I can't explain what that means for others, such a statement plays out for me in the form of ideas coming to mind that are creative, worthwhile, and unequivocal. In that context, God said to me, "I gave you a brain. Use it in my defense! I know I can stand the test!"

Every mathematician has a streak of philosopher in him. This is seen historically in the lives and works of people like Rene Descartes, Sir Isaac Newton, Blaise Pascal, Johannes Kepler, Bertrand Russell, and Alfred North Whitehead. These men were not all theists, but their work exemplifies mathematicians making connections between philosophy and/or religion and mathematics.

My life has been dedicated to writing mathematics textbooks for college students, and if God has truly blessed me with a talent, it seems to be this. One of my goals is to enhance my textbook writing by the inclusion of more applications, a magnificent obsession that is both unavoidable and heartfelt. No matter what I am involved in, whether eating Wendy's hamburgers, going bowling, playing golf, playing baseball, or sitting in church, I seem to find mathematics. That trend recurred in a most unexpected way, following my heart attack in September of 1998. It is a continuing story.

As I stated in *One Man's Journey Through Mathematics,*

I have always tried to inspire my sons by teaching them and talking to them about mathematics, in hopes that they might pursue or at least be well versed in the subject. In the case of my oldest son, Lowell, it worked because is he now an actuary with Conseco, here in Carmel, IN. In the case of Chris I remember one incident while the family was riding in the car, and I saw a road sign which suggested some kind of application. I said enthusiastically, "Chris, there is a mathematical formula for that application."

> Chris, somewhat weary of such stories responded, "Dad,
> you have a formula for everything!" Needless to say, one's
> parental attempts sometimes go awry.

September 23, 1998 is my heart attack "birthday." The evening
of that day, my precious wife Elaine and I went for a walk. I became
aware of a feeling of tightness in my chest. It bore little resemblance
to what I thought chest pains should be; it was not like being hit with
a baseball bat or having a cement block placed on my chest. Instead, I
seemed to want to push my chest out to relieve the tightness. Nothing
more. When we got home, the tightness went away.

I went to bed and slept well until I got up about 3:30 AM to go to
the bathroom. The chest tightness was now back, but this time I was
in a cold sweat. Something was wrong! I woke Elaine and told her
that we should drive to the ER. In hindsight, we should have called an
ambulance. Being her usual law-abiding self, Elaine was stopping at
the red lights. I finally told her to stop, look left and right, and then run
the lights. I knew my condition was worsening.

To our amazement, when we got to the ER, there were no other
patients—I had the attention of all the doctors and nurses. They infused
IVs and connected me to an EKG. Shortly, the doctor patted me on the
shoulder and said, "Well, I have bad news, and good news! The bad
news is that you are having a heart attack. The good news is that it is on
the right side, and that gives you a better chance for recovery." I looked
over at Elaine sitting nearby and pointed my finger up, and she did as
well. The unstated implication was that we were both praying.

My life really did flash before my eyes. I thought of all the times
I had resisted God's will, pursued my passions instead of His, and
thought I was always going to be in control. My prayer went something
like this: "God, you finally have my attention! I know you are in control of whatever happens from here. If you want to take me, I'll accept
that. If you want me to live, it is my preference to stay. I would love to
see my grandchildren (at that time none existed)!"

At that time, a certain peace came over me that I can never explain.
Elaine commented later that in similar situations, I would normally
panic—but not this time. I think it was because I had turned everything

over to God. It was not 30 seconds later that I went into cardiac arrest: *I coded.* I have an absolutely clear memory of that experience, at least until I passed out. I soon returned to consciousness, hearing the doctor casually say, "Well, he's back now." He had resuscitated me with the defibrillating paddles. I later saw the rectangular burn marks on my chest—even under that stress, I managed to discover some math. I know that may sound bizarre to a non-mathematician. It is a wonder I didn't take their measurements and compute the area and perimeter.

I left the hospital on September 27 after receiving two stents for the blocked arteries on the right side of my heart. It was a real blessing to be treated with the stents instead of undergoing bypass surgery. It is a paradoxical, or ironic, fluke of probability that I had been exercising for several years, had been watching my diet fairly well, and had even passed a treadmill test with flying colors only five weeks before. My blockage was some combination of bad genetics and poor eating habits from my youth. (As I think back, I am not surprised to have clogged arteries, considering the meals we ate. I recall meals consisting entirely of potatoes fried in bacon grease.)

Following my recovery, I asked God what He wanted from me for the rest of my life. It was at Jubilee Lanes before my bowling league, when the paradoxical message came through: "Marv, look at all the passions you have had in your life—your sports, bowling, softball, baseball, Purdue University athletics, baseball camps, hiking in Utah, efforts at career advancement, and moving up the academic ladder, all of a self-centered nature. Why don't you give me the same passion?" I heard the message. It initiated intense reading of the Bible and other theological materials. I had never read the Bible, rebelling against the thees, thous, and thus-sayeths of the King James Version in my youth. Now it was like having a straw in a milkshake of wisdom. I couldn't slurp up the knowledge fast enough. I read the Bible six times, but the more readable *New Living Translation*, or *The Message* by Eugene Peterson.

I could not believe what I was finding. You have one guess. It has to be—*mathematics.* Could God give me spiritual wisdom and mathematics at the same time? I couldn't be that blessed! The first mathematics I saw was frequent references to the number seven. It turns out that the word *seven* or *seventh* is mentioned in the first 21 out of 39 books of the Old Testament. The first occurrence is the second chapter of Genesis, and the last appearance is in the next-to-last chapter of the New Testament book of Revelation.

Descriptions of Christianity and Faith

It is appropriate to define the concepts *Christian* and *faith*. The notion of a *Christian* is not clearly defined. My pastor, Dave Rodriguez, asserts that in the U.S., many people who do not already identify themselves as participating in some other religion such as Islam or Judaism somehow automatically classify themselves as "Christian." They feel this way despite never attending a church, practicing in any aspects of Christian faith, or having any relationship with the Christ of Christianity.

Keeping in mind the diversity of my audience, I feel compelled to consider some related definitions, although delving into such definitions in depth is beyond the focus of this book.

A *Christian* is a follower of Christ according to the books of the New Testament. One becomes a *follower of Christ* by turning around from a life of self-centeredness and moving toward the perfection modeled by Christ. This is accomplished by an act of faith.

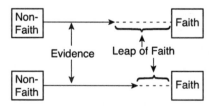

Figure I.1 The act of faith

According to Hebrews 11:1, "*Faith* is the confident assurance that what we hope for is going to happen. It is the evidence of things we cannot see." Coming to faith is a decision of the *mind*, *heart*, and *will*—the faith equation Faith = (Mind) + (Heart) + (Will). Frederick

Buechner writes, "Faith is homesickness. Faith is a lump in the throat. Faith is less a position on than a movement toward, less a sure thing than a hunch. Faith is waiting. Faith is journeying through space and through time." I have come to believe that *faith* is wading through a chain of evidence and reaching a reasoned decision to become a follower of Christ.

According to Lee Strobel, "I see faith as being a reasonable step in the same direction that the evidence is pointing. In other words, faith goes beyond merely acknowledging that the facts of science [*mathematics*] and history point toward God. It's responding to those facts by investing trust in God—a step that's fully warranted due to the supporting evidence."

J. P. Moreland writes, "… faith is a power or skill to act in accordance with the nature of the kingdom of God, a trust in what we have reason to believe is true. Understood in this way, we see that faith is built on reason. We should have good reasons for thinking that Christianity is true before we dedicate ourselves completely to it."

Biblical Translations and Interpretations

When you hear a sermon, keep in mind that there is always an element of *interpretation* involved in reading and quoting the Bible. How many sermons have you heard on Psalm 23, John 3:16, or the Sermon on the Mount? In addition, there have been more than 3,000 translations of the Bible from the Hebrew and Greek; and although Christians have the belief that in its original form the Bible was inspired by God, it was still written down by humans, and humans make interpretations as they translate.

Let's consider two illustrations of Biblical interpretation. The first is provided by John Eldredge in his book *Waking the Dead*, when he takes issue with a passage in Jeremiah that refers to God as "the Lord Almighty." (See Jeremiah 7:3.) Eldredge went to an "original" Hebrew translation and found that the passage means, "the God of angel armies," which he interpreted on his own to mean, "The God who is at war." Eldredge wanted to make the point that God is on Earth with us fighting for our hearts. I totally agree with the end result of Eldredge's pursuit of knowledge. His illustration brings home my point that *interpretation* is an aspect of the study of the Bible. I used the word "original" in quotes regarding the Bible, because there are 5,366 ancient manuscripts considered to be 99.5% identical, but none is the original. At this point in history, the original does not exist.

A second example of interpretation is found by considering the same passage in the Bible, but in different English translations. You can do this easily on the free website www.biblegateway.com. The following is my *favorite* Bible passage, but it is only my favorite when read in the New Living Translation, 1st edition:

> **Psalm 92:10** *(New Living Translation; 1st ed., 1996): But you have made me as strong as a wild bull. How refreshed I am by your power!*

Wow! That quote, in this translation, is empowering and encouraging to me. Compare this same passage in five other translations of the Bible:

> **Psalm 92:10** *(King James Version): But my horn shalt thou exalt like the horn of an unicorn: I shall be anointed with fresh oil.*

> **Psalm 92:10** *(English Standard Version): But you have exalted my horn like that of the wild ox; you have poured over me fresh oil.*

> **Psalm 92:10** *(New International Reader's Version): You have made me as strong as a wild ox. You have poured the finest olive oil on me.*

> **Psalm 92:10** *(New International Version—UK): You have exalted my horn like that of a wild ox; fine oils have been poured upon me.*

Psalm 92:10 *(The Message): But you've made me strong as a charging bison, you've honored me with a festive parade.*

The latter five translations leave me unaffected and uninspired. But do I believe God speaks to me through the verse of the New Living Translation (NLT), 1st edition? Yes: *That message is creative, worthwhile, and unequivocal,* empowering a guy like me who has struggled with his self-esteem from childhood to the present. These varied translations bring home the point that *interpretation* is an aspect of the study of the Bible. Now, look at the most recent version of Psalm 92:10, as published in the second edition of the NLT:

Psalm 92:10 *(New Living Translation; 2nd ed., 2004): But you have made me as strong as a wild ox. You have anointed me with the finest oil.*

In this form, I must confess that the empowerment and encouragement I found in the 1st edition are gone. When I spoke to the editors of the NLT from Tyndale House publishers, they commented that the style of the 1st edition of Psalms was written more in prose, but in the 2nd edition they wanted to return to a poetry style, as David is purported to have written.

I now invite you to join me on a journey of insight into issues of the Christian faith as presented in a framework of mathematics. Bring your mathematical background, your theological knowledge, your Bible, and your imagination, and enjoy the ride.

The website for this book, www.thefaithequation.com, provides discussion and research questions, which are also at the end of this book, and links to more math depth, a blog, and numerous quotes. This book and the discussion questions provide opportunities for a capstone course for seniors majoring in mathematics or the sciences. Check the website on an ongoing basis for new questions and updates.

Marvin L. Bittinger
January 2007
Indianapolis, Indiana

Chapter 1

Apologetics and Faith Axioms

We need to admit the mind into Christian fellowship again. We need the mind disciplined in Christ, enlightened by faith, passionate for God and His creation, to be let loose in the world.

—J.P. Moreland

It seems as though Christians today either question too much or not enough. They either blindly follow what's said in the pulpit or constantly search for evidence. And although many would say that searching for evidence takes away from faith, many of us who want more learned evidence find that this search brings us closer to God—closer to an understanding that the universe couldn't work without God. The search brings us closer to explaining God in terms we can understand and relate to and ultimately, as scientists, teachers, and educated people, in terms in which we can find evidence.

The idea that we can ultimately use logic to reason about our faith is based on a concept called *apologetics*. Apologetics, as applied to a faith, is the "reasoned defense," or "the science and art of defending a faith." The word *apologetics* comes from the Greek word Απολογια, which is pronounced "apologia" and means *defense*. The New Testament was written in Greek. When a lawyer in Greece gave a defense in a trial, the lawyer was presenting an *apology* or a reasoned defense or explanation. The meaning is different than our word *apology* today, but the word's origins are in the same root. You could try to give a reasoned defense, or apologetic argument, for any religious faith. In this book, the word refers only to Christian apologetics.

A person who pursues an apologetic argument is called an *apologist*. You can think of an apologist as a lawyer in a courtroom arguing a case for Christianity, linking one piece of evidence to the next to create a chain of proof. When a chain of proof leads to evidence beyond a reasonable doubt, then a case for Christianity exists. Noted scholar and apologist William Lane Craig once said that to him, such a chain of evidence does exist for coming to Christian faith, but few people are aware of it even in university circles, and even fewer make an effort to seek apologetic understanding. Many people like myself have had faith imposed upon them by family, friends, or a podium-pounding minister.

Many types of evidence are used in apologetics; among these are historical evidence, eyewitness evidence, documentary evidence, corroborating evidence, scientific evidence, rebuttal evidence, and mathematical evidence. By way of exemplifying the notion of apologetics, we'll consider an example using historical evidence. After this chapter, we'll primarily use evidence in a framework of mathematics.

Craig says the following:

The goal of historical knowledge is to obtain probability, not mathematical certainty. An item can be regarded as a piece of historical knowledge when it is related to the evidence in such a way that a reasonable person ought to accept it. This is the situation with all of our inductive knowledge: we accept what has sufficient evidence to render it probable. Similarly, in a court of law, the verdict is awarded to the case that is made most probable by the evidence. The jury is asked to decide if the accused is guilty—not beyond all doubt, which is impossible—but beyond all reasonable doubt. It is exactly the same in history: we should accept the hypothesis that provides the most probable explanation of the evidence.

One Apologetic Argument: An Example

One kind of apologetic argument is based on historical evidence or authenticity (historicity). For example, what if we want to create an apologetic case for the existence of Jesus Christ? We begin by researching documents and history to create an argument by considering the existence of a different person from history, Alexander the Great. If we reasonably trust the accuracy of historical documents and the veracity of the arguments that Alexander the Great existed, then by analogy similar documents and knowledge provide evidence for the existence of another historical figure, Jesus of Nazareth.

To cite historical evidence regarding Alexander the Great, who lived from 356–323 BC, I found a myriad of historical writings, but I'll use the following two books:

- *The Romance.* An early version of the book was composed in Alexandria after Alexander's death; the work was edited, retold, and extended during many centuries, the last original being a publication in Venice as late as 1529 AD. The oldest surviving manuscript is from the third century AD, and a modern translation by Richard Stoneman is available. The book has been attributed to Calisthenes, but he died in 327 BC. The book covers Alexander's life, including many fantastic tales that we now know never happened.
- *The Library of History, Book XVII.* This book, by Diodorus Siculus, was written sometime after 50 BC. Book XVII is available in a modern English translation by C. B. Welles. Diodorus, a historian living in Sicily from 80 BC to 20 BC, was known as an uncritical author who used reliable source material.

Other books that still exist have publication dates around 40 AD, 100 AD, 140 AD, and 200 AD. Most of these existing publications reference other books that were lost in antiquity. Keep in mind that in this example, we're trying to establish the existence of Alexander to provide one piece of an analogy. Hundreds if not thousands of other texts regale us with tales—some true and some unconfirmable—of Alexander's life. Although we can provide evidence that to some extent substantiates the authenticity of the texts, we can't prove the authenticity of many events. (Did he cut the Gordian knot? Myth says yes, but we just don't know.) Some texts claiming to have details have been proven unreliable by scholars. However, a reasonable person would *at least* believe in the existence of Alexander the Great.

> It should be noted that in the historical and academic communities, BC (Before Christ) is now considered politically incorrect. They now use BCE (Before the Common Era). This book focuses on Christianity, so I choose to use BC and AD.

Living well before the birth of Christ, Alexander was the King of Macedonia from 336–323 BC and military conqueror of Asia Minor, Syria, Egypt, Babylonia, and Persia. His reign marked the beginning of the Hellenistic civilization, which spread through the Mediterranean and Middle East and into Asia after Alexander's conquests. Although the Greek city-states stagnated, their culture flourished elsewhere, notably at the city of Alexandria, on the coast of the Mediterranean Sea in Egypt.

Alexander is said to have marked off the streets of Alexandria before moving on to other conquests. A great library was eventually established from which the city's influence on art, letters, commerce, and mathematics was so great that the era is sometimes called the Alexandrian age. According to Victor Katz in *A History of Mathematics*, "The most important mathematical text of Greek times, and probably of all time, the *Elements* of Euclid, written about 2,300 years ago, has appeared in more editions than any other work other than the Bible.... [I]t is generally assumed that Euclid taught and wrote at the Museum and Library at Alexandria." In addition to Euclid, this age spawned many other great mathematicians, including Archimedes, Appollonius, Eratosthenes, and Hipparchus, whose work in astronomy eventually led to that of Ptolemy, Copernicus, Brahe, and Kepler. Although Alexander was not a noted mathematician, he provided a foundation for immense growth in mathematics.

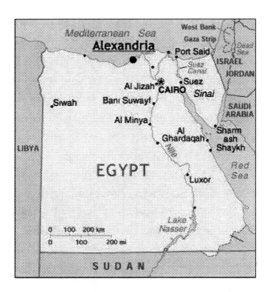

Figure 1.1 Historic Alexandria on the coast of Egypt

The historical evidence beyond a reasonable doubt demands the verdict that Alexander the Great lived. He existed. And although many historians and scholars could argue endlessly that he may or may not have been responsible for certain contributions, there's a reasonable abundance of evidence that he was a great man and responsible for a great deal, even though the earliest known manuscript we can check dates to 350 years after his death. *We have presented an apologetic argument for the existence and some contributions of Alexander the Great.*

Alexander is also credited with the spread of the Greek language throughout much of the world at that time. This provided a common language when Christ's disciples began to travel and share their faith. The New Testament was also written in Greek.

To create an apologetic argument for the existence of Christ, let's make an analogy to the evidence provided by one author, Lee Strobel, once an attorney and now an author, in his book, *The Case For Christ.*

In citing Biblical scholar Craig Blomberg, Strobel refers to the Gospels of Matthew, Mark, Luke, and John as follows: "The standard scholarly dating, even in liberal circles, is Mark in the 70s [A.D.], Matthew and Luke in the 80s, John in the 90s." Then, by analogy to the historical argument for Alexander the Great, whose writings can be checked to only 350 years after his death, why would anyone, in particular a mathematician, question the Biblical writings regarding Christ's existence, which occurred less than 90 years after His death?

Further historical evidence of the existence of Christ is found by consulting non-Christian historians. For example, Flavius Josephus (37 AD to about 100 AD), a first-century Jewish historian who was not a Christian, wrote of Jesus and the Christians in his book *The Antiquities of the Jews*; he referred to Jesus as "a wise man, a doer of wonderful works" who "drew over to him both many of the Jews and many of the Gentiles." Other historians such as Cornelius Tacitus (56 AD–117 AD); Thallus, a Samaritan historian (ca 52 AD); and Suetonius in 120 AD wrote about the existence of Jesus.

We have established an apologetic argument for the existence of Christ. Although a great deal more evidence can be considered, it is not our goal to further pursue historical evidence for the existence of Christ because doing so takes us away from a mathematical framework for our apologetics.

An Apologetic Argument for the Reliability of the Bible

Let's consider another example of an apologetic defense. In this case, we consider the reliability of the Bible. One of the strongest claims for the historic reliability of the Bible comes from manuscript evidence. According to Gary Habermas in the book *Why I Am a Christian: Leading Thinkers Explain Why They Believe,* "Generally, several qualities enhance manuscript value, assisting textual scholars in arriving at the best reading of the original text. The strongest case is made when many manuscripts are available, as close in time to the original autographs as possible." The following chart from Bruce Metzger in his book *The Text of the New Testaments* allows a comparison of the New Testament with other ancient works.

There are 5,366 ancient texts that have been found 99.5 percent free of textual discrepancies. If you include ancient translations in languages other than Greek, the number of surviving ancient manuscripts extends to 24,970 copies. Do people challenge the textual accuracy and integrity of the *Iliad,* when it has 643 ancient manuscripts? It is well studied in schools and colleges. Why do some challenge the general accuracy and integrity of the New Testament (nonfiction) when it has 24,970 ancient manuscripts, of which 5,366 ancient texts have been found 99.5 free of textual discrepancies? Just as in the *Iliad*, a fictional work, individual details in the New Testament can be debated. The *Iliad*, after all, is based on an oral tradition, and we have only a rough date for its writing. With the New Testament, we have a bit more detail about its time period and authors. These details can be debated, but they don't detract from the overall fact, proven by our analogy argument: The New Testament can be considered for the most part accurate.

Author	Book	Date Written	Earliest Copies	Time Gap	# Of Copies
Homer	*Iliad*	800 BC	c. 400 BC	c. 400 yrs	643
Herodotus	*History*	480–425 BC	c. 900 AD	c. 1,350 yrs	8
Thucydides	*History*	460–400 BC	c. 900 AD	c. 1,300 yrs	8
Demosthenes		300 BC	c. 1100 AD	c. 1,400 yrs	200
Plato		400 BC	c. 900 AD	c. 1,300 yrs	7
Caesar	*Gallic Wars*	100–44 BC	c. 900 AD	c. 1,000 yrs	10
Livy	*History of Rome*	59 BC–17 AD	4th century (partial) mostly 10th century	c. 400 yrs c. 1,000 yrs	1 partial, 19 copies
Tacitus	*Annals*	100 AD	c. 1100 AD	c. 1,000 yrs	20
Pliny Secundus	*Natural History*	61–113 AD	c. 850 AD	c. 750 yrs	7
New Testament (Greek Translations)		50–100 AD	c. 114 AD (fragment) c. 200 AD (books) c. 250 AD (most of NT) c. 325 AD (complete NT)	+ 50 yrs 100 yrs 225 yrs	5,366

More Apologetics: Could a Lie Have Been Perpetrated?

Some who wish to negate the historical authenticity (historicity) of the New Testament challenge the lives of the 11 disciples and the apostle Paul who remained after his death, asserting that they lied about Christ's life, words, and resurrection. Let's suppose that these men did attempt to perpetrate such a lie. What would have been the incentives or consequences for them?

Assuming you accept the reliability of the New Testament, let's look at Matthew 28. Matthew was the first of the four Gospels written and could be presumed to be most accurate. In it, we're told that on the Sunday after Christ's crucifixion, Mary Magdalene and possibly other women went to the tomb. They discover the stone displaced from the tomb and no body remaining. The discoverer(s) of the empty tomb was a *woman* or *women* depending on which gospel you read. In the society of that day, women were held in very low regard. If the disciples were attempting a hoax, wouldn't they have sent men to discover the empty tomb? The men would have commanded a higher level of believability.

The disciples were committed to the Ten Commandments, one of which says not to tell a lie. It would have taken a real stretch of truth to convince the world about an empty tomb unless one actually existed. In addition, it would have been difficult for all 12 of them to lie as a group and keep the secret. Look at the lives of the disciples and the Apostle Paul after Christ's resurrection. These men went on to spread the message of Christ. Many of them suffered extreme persecution and tragic deaths. Andrew was crucified, James was beheaded, Philip was martyred, Bartholomew was skinned alive and beheaded, Thomas was speared to death, Peter was crucified upside down, and Paul was beheaded. Would all of these men have endured these sacrifices for the sake of a lie? I think at least one of them would have bailed out.

In addressing the issue of the 11 disciples and Paul lying about Christ's life, words, and resurrection, William Lane Craig in his book *Reasonable Faith* writes regarding these men:

…they either sincerely believed that this religion was true or they did not. If not, then they would never have chosen it for their own and rejected the safer, more customary religions. But if they believed it to be true, then the resurrection of Jesus cannot be avoided. For had he not risen, contrary to his prediction, that would have destroyed the very foundation of any faith the disciples had. Moreover, their own religion prohibited lying and any bearing of false witness. And besides this, no one, and especially so many, would be willing to die for a lie that they themselves had made up, a lie that would bring them absolutely no worldly good. And it is clear from the writings that the apostles were not madmen. Finally, the conversion of the apostle Paul bore witness to the reality of the resurrection.

Note Craig's use of logic. In math, something is either true or it is not. Even as we look at probabilities later in this book, we use mathematical logic to make our case. The answer is yes or no, true or false.

Let's contrast the preceding historical argument, or case for Christ, with one for the first landing on the moon, by Neil A. Armstrong on July 20, 1969. It has been purported by some that the landing was a hoax. An excellent article refuting such allegations is found in http://science.nasa.gov/headlines/y2001/ast23feb_2.htm. However, scientists, engineers, astronauts, NASA, and millions of people around the world know that we did make it to the moon and land there just as history attests. Think of the thousands of people involved in the development of the space program, and of the lives of the astronauts themselves. If the moon landing were indeed a lie, wouldn't one of them have spilled the beans in the more than 38 years since the event? We all know that as the number of people involved in keeping a secret increases, so does the likelihood of someone confessing. Unless the conspiracy was confined to an incredibly small group of people (which is all but impossible), a lot of people would have had plenty of incentive to tell what they know—especially now, nearly 40 years after the fact.

The historical evidence (historicity) for the landing on the moon has a probability of 1.0 in my mind. For the existence and life of Christ, the evidence convinces me that the probability of the existence and life of Christ is about 0.99999999. How does a leap of faith come into play? It comes mathematically, by moving from 0.99999999 to 1.0.

Recall the faith equation

Faith = (Mind) + (Heart) + (Will)

We have brought evidence to bear to strengthen the *mind's* ability to accept Christ. Admittedly, we have only touched on the evidence in this chapter, but as the book progresses, the evidence will become stronger. There is a great deal more evidence that can be considered regarding the historical reliability of the New Testament. The works of authors like Lee Strobel, Josh McDowell, and William Lane Craig are just three out of a vast amount of literature on the subject. For example, Strobel discusses other eyewitness evidence, as well as documentary evidence, corroborating evidence, scientific evidence, rebuttal evidence, and so on.

Although there are occurrences of numerical information in our preceding discussion, examining historical evidence further takes us far beyond our focus on a mathematical framework for apologetics. The intent in the preceding was to exemplify how an apologetic argument can proceed when it isn't particularly mathematical. Now, let's get into some mathematics!

Mathematical and Faith Axioms

We have considered some examples of how an apologetic argument works and what it is. The thrust of this book is to use apologetics more or less focused on the Bible. As a mathematician, this helps authenticate what I have already accepted through faith to be true. The Bible is the common foundation for all Christianity movements, but many differences in faith among those movements have little to do with the Bible—religious articles of faith aren't always Bible based.

That said, I am left with my faith and what I have accepted as fact and my ability as a mathematician and scientist to provide apologetic arguments. We are strengthening the *Mind* part of the faith equation. It is the goal of this book to give many reasons based on mathematics for becoming a follower of Christ.

We will draw a comparison between forming axioms and theorems in mathematics to forming "faith axioms" in Christianity, although little of the standard deductive proof of mathematics will prevail. People often ponder what mathematicians do. Simply stated, they make assumptions or axioms (as few as possible), and then they try to prove, using rules of mathematical logic, as many theorems as they can from the axioms. But how do they arrive at the initial axioms? The answer is multifaceted, but here is how such a process might work. Consider the tip of a needle or a dot on a piece of paper, and think of it as a *point*. Now, think of connecting two dots by drawing a line between them along a ruler or using a piece of string. From those practical experiences, we can come to this geometric axiom:

> *Axiom 1.* Given two distinct points, there exists one and
> only one line containing the points.

Using other axioms as well as this one, a mathematician might go on to prove a theorem like the following:

> *Theorem 1.* A line contains an infinite number of points.

Granted, a definition of *infinite* is also needed before a proof can be completed. We can't take the space in this book to delve into details of such a proof. You can consult a book on advanced geometry such as Moise's *Elementary Geometry From an Advanced Standpoint*, or the foundations of the real number system such as Chartrand, Polimeni, and Zhang's book *Mathematical Proofs: A Transition to Advanced Mathematics* for more detail.

Mathematicians typically operate in the context of a mathematical system. Stated briefly, a *mathematical system* starts with a universal set of objects that is the framework of the system. Examples are the real number system in algebra and the points of a plane in geometry.

A mathematical system also contains a set of definitions; for example, in the real number system we can define an *even integer* as an integer that can be expressed as a multiple of 2. For example, 2, 4, 8, and 364 are even; but 3, 5, 7, and 365 are not. Another definition is of the symbol $\sqrt{a}$, which is the non-negative square root of a non-negative number, and is that non-negative number which when multiplied by itself gives a. For example, $\sqrt{9} = 3$ and $\sqrt{289} = 17$.

We also have an underlying set theory and a set of relations and/or operations. In the real number system, *less than* is a relation: For example, $3 < 5$. *Addition* is an operation: For example, $3 + 5 = 8$. A mathematical system also uses a set of logical axioms that are the underlying rules of logic and reasoning. For example, if we know that each of the sentences "If it is raining, then it is wet outside" and "It is raining" are true, then we can deduce that "It is wet outside" is true.

There is also a set of nonlogical axioms; these axioms pertain to the elements, relations, and operations—the entities most often studied by mathematicians. In most discourse, these are referred to as the *axioms*. From the axioms, we use the rules of logic to deduce theorems. In effect, we create a sequence of statements deduced by the logic, and the last statement in the sequence is the theorem.

In the mathematical system known as the real numbers, as it pertains to ordinary algebra, we can use the axiom known as the distributive law, $a(b + c) = ab + ac$, to prove the theorem

$$(x - 3)(x + 5) = x^2 + 2x - 15$$

A definition in the real numbers may be that of a rational number or an irrational number. A *rational* number is a number that can be expressed as a ratio of two integers. An *irrational* number cannot be expressed as a ratio of two integers. For example, the numbers $\frac{2}{3}$, $-\frac{7}{8}$, and 6.75 are all rational, but $\sqrt{2}$, π, and $-\sqrt[3]{55}$ are not rational: They are irrational. That a number like $\sqrt{2}$ is irrational is a theorem whose proof is too lengthy to consider here. See Chartrand, Polimeni, and Zhang's *Mathematical Proofs* for a detailed proof.

Let's look at the idea of *parallelism* for another example. In mathematics, if you change the axioms, you change the consequences. We'll use geometry as a medium to illustrate the effect of the change of just one axiom. Many readers will remember taking geometry classes and working on endless proofs. Without revisiting a long list of axioms and definitions, let's just consider one: parallelism. First we have a definition, and then we have an axiom:

> *Definition.* The lines L_1 and L_2 are parallel if they lie in
> the same plane and do not intersect.

In our normal everyday usage of geometry, we also think of parallelism as lines that are always the same distance apart or that do not cross.

Let's consider an axiom that is pertinent in a mathematical system known as *Euclidean Geometry:*

> *Math axiom: Euclidean Parallel Postulate (EPP).* Given a
> line L and a point P not on the line, there is one and only
> one line M through P parallel to L.

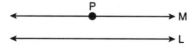

Figure 1.2 Two parallel lines do not intersect.

Using the EPP axiom, we can prove the following basic theorem. Such a proof is lengthy and will not be considered here. You can find the proof in Moise's book, noted earlier:

> *Theorem (in Euclidean Geometry).* The sum of the angles
> of a triangle is 180°.

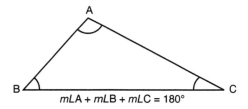

$$mLA + mLB + mLC = 180°$$

Figure 1.3 A Euclidean triangle

This result might be important to an engineer in constructing a building or a bridge. But suppose we change that axiom to one that belongs to a non-Euclidean mathematical system know as *Riemannian Geometry*:

> *Math axiom: Riemannian Parallel Postulate (RPP).* Given a line *L* and a point *P* not on the line, there are no lines *M* through *P* parallel to *L*.

In the following figure, you can see the line *L* and the point *P*, but imagine the possibility that it is impossible to have a line parallel to *L*.

Figure 1.4 No intersection

Although it's difficult to imagine, there does exist a mathematical system called Riemannian Geometry in which this axiom is accepted and from which there are practical consequences. Using the RPP axiom, we can prove the following rather surprising theorem:

> *Theorem (in Riemannian Geometry).* The sum of the angles of a triangle is greater than 180°.

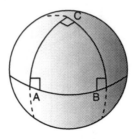

Figure 1.5 A globe or sphere

There are situations where this theorem is relevant: for example, our globe. Can you really fly on a straight line from Indianapolis to Los Angeles? Not really. Say that a plane can fly on a single compass heading from one city to the other; the path the plane follows is more like a great circle—a circle whose center is at the center of the earth. In spherical geometry, lines are arcs of great circles, and the sum of the angles of a triangle exceeds 180°. If you wanted to travel to LA in a straight line, you would have to install a drill bit in the nose of the airplane and bore through the earth.

We reach a conclusion in mathematics based on the two very different theorems—different axioms give different consequences. Think of Euclidean Geometry as having application to a smaller realm like the construction of a bridge or building. Think of Riemannian Geometry as having application to a larger realm like air or space travel.

Let's contrast math axioms with faith axioms.

Math axioms are assumptions that have intellectual consequences and sometimes-practical consequences to applications like building bridges or flying from city to city. This book covers both math and faith. I've found that along with math axioms that support theorems in math applied to real life, we also arrive at faith axioms that apply to the spiritual side of life. *Faith axioms* are assumptions that have intellectual consequences to the spiritual side of your life: for example, *There exists exactly one God*, or *Love others as oneself*, or *God is the Designer and Creator of the Universe*.

When you change math axioms, you change the consequences. When you change faith axioms, you change the consequences. Consider *There exists exactly one God*. The Christian religion has one and only one God. The Hindu religion has over 300,000 gods. In many first-world countries, the faith axiom *There exists exactly one God* changes as we pursue multiple gods of success, pleasure, sexual conquest, power, and riches. The consequences of the pursuits of these multiple gods take away from love and the passion for a life plan that God (one and only one) puts in our heart.

The Journey to Faith Axioms

Forming faith axioms is a journey through all your life experiences, reading, studying, scientific (in our case, mathematical) knowledge, a pursuit of apologetics, Bible examination, prayer, and so on. The road to trust in a faith axiom is the resolution of a dynamic tension between *belief* and *questioning*. Ultimately, you *choose* the faith axioms you trust and follow. I prefer the word *questioning* to the more theologically accepted *doubt.* On some issues, the dynamic tension between *belief* and *questioning* may always exist, at least until we leave these dimensions.

Forming faith axioms is a progression through an alternating rhythm of taking hold and letting go, a believing and a questioning. Such thinking can be thought of as *true skepticism.* In an article in *Christianity Today*, Mark Buchanan asserts, "Any skeptic worthy of the name is both hunter and detective, stalking the evidence, laying ambush, rummaging for clues…. Skeptics are passionate about finding truth out." If this were what it means to be a true skeptic, then I would call myself a true skeptic establishing my faith axioms in Christianity.

Emily Dickinson once said, "We both believe, and disbelieve (doubt) [question] a hundred times an hour, which keeps believing nimble." The Biblical passage in Mark 9:23-24 also addresses this idea: "…I do believe, but help me not to doubt." As a mathematician and a Christian, one of my favorite mottos is "I do believe in the Christian faith, but I must continue to ask questions." Another is "I don't have all the answers, but I have most of the questions."

A Faith Axiom on the Bible

Biblical *inerrancy* or *infallibility* is a theological and personal belief on the part of many Christians that the Bible in its original form is inspired by God and *without error*. Much of the controversy boils down to coming to grips with a definition of the word *errancy* in a Biblical context. What may seem like errors can be resolved by considering issues in the context of the culture and scientific knowledge of the times, or by realizing that a topic is a parable, is symbolic, is an allegory, or is a metaphor. A certain distance or mystery exists between what words say and the realities they represent.

The issue of Biblical inerrancy is a bottomless pit of controversy among Christians and secular scholars. At the least, such a topic can't be considered in a mathematical framework, although the subsequent brief discussion includes some numerical aspects. At the most, inerrancy is too involved to cover thoroughly in this book, but it plays a critical part in one of my personal faith axioms—that I believe the Bible is the most extraordinary way to the character of God. Lindsell's *The Battle for the Bible* and Ehrman's *Misquoting Jesus* both tackle this difficult topic in detail, although each has a widely differing viewpoint. For purposes of establishing a faith axiom regarding the Bible, and to be forthright to you about a faith axiom I hold about the Bible, I feel compelled amid the controversy to give some views on Biblical inerrancy and interpretation. We are considering the dynamic tension between belief and questioning.

Copying Issues

You must always keep in mind that although most Christians assume the Bible in its original form was inspired by God, fallible men wrote down, copied, and recopied the information, mainly in Hebrew for the Old Testament and Hebrew and Greek for the New Testament. If you and I lived in those times, and I wanted to prepare a copy of my Bible for you, I would have to employ a copyist to attempt a handwritten duplicate. To make an exact duplicate would be a daunting task. From my experience, publishing a math textbook error-free even in this day and age is the most challenging if not impossible part of writing. Considering that there exist copies of copies of copies of copies of Biblical manuscripts, isn't it reasonable that variants or errors may have occurred? Let's consider some of these situations.

My wife and I recently visited Trinity College in Cambridge, England. There we saw, firsthand, an eighth-century handwritten copy of the New Testament—it was 800+ years old. On the page we viewed, there was a hole with a word or part of a word missing. Would that missing word change the meaning of a Biblical message as that particular copy was copied and recopied down through history? And how many copying errors may have occurred during the prior 800 years? I have no definitive answer to these questions, but they illustrate how errors can be made and passed on to other generations. Ehrman asserts that "It is one thing to say that the originals were inspired [by God] but the reality is that we do not have the originals—so saying they were inspired doesn't help me much, unless I can reconstruct the originals.... What we have are copies made many *centuries* later. And these copies all differ from one another, in many thousands of places."

Changes by a Copyist or Translator

Various authors have retranslated sections of the Bible over the years. Even the four Gospels have four versions written by different authors at different times and even in different languages. Owing to translation errors in words or phrases that may not mean exactly the same thing, changes have also been made reflecting the philosophical, theological, and political views of the copier as well as personal attempts to improve the writing. Even in modern times, kings and corporations have issued new—and sometimes radically altered—translations of the Bible. Ehrman cites the results of variants in scripture regarding the Lord's Prayer as discussed in Parker's book, *Living Text of the Gospel.* Compare the following two texts of the Lord's Prayer as given in Luke and in Matthew:

Luke 11:2-4 (English Standard Version): 2 *And he said to them, "When you pray, say: 'Father, hallowed be your name. Your kingdom come.* 3 *Give us each day our daily bread,* 4 *and forgive us our sins, for we ourselves forgive everyone who is indebted to us. And lead us not into temptation.'"*

Matt 6:9-13 (English Standard Version): 9 *Pray then like this: "Our Father in heaven, hallowed be your name.* 10 *Your kingdom come, your will be done, on earth as it is in heaven.* 11 *Give us this day our daily bread,* 12 *and forgive us our debts, as we also have forgiven our debtors.* 13 *And lead us not into temptation, but deliver us from evil."*

The quote from Matthew contains nine extra words. According to Ehrman, some manuscripts add additional phrases from the passage in Luke to the Lord's Prayer so that it reads as in Matthew 6:9-13. He comments that "This scribal tendency to 'harmonize' passages in the Gospels scripture is ubiquitous"—this means it happens all through the Bible.

I found further evidence as I examined Luke 11:2-4 in the New Living Translation. Note in particular the footnote [a]:

Luke 11:2-4 (New Living Translation): 2 *Jesus said, "This is how you should pray: [a] "Father, may your name be kept holy. May your Kingdom come soon.* 3 *Give us each day the food we need, [b]* 4 *and forgive us our sins, as we forgive those who sin against us. And don't let us yield to temptation. [c]"*

Footnotes: *[a] Luke 11:2 Some manuscripts add additional phrases from the Lord's Prayer as it reads in Matt 6:9-13. [b] Luke 11:3 Or Give us each day our food for the day; or Give us each day our food for tomorrow. [c] Luke 11:4 Or And keep us from being tested.*

Issues of Punctuation and Spacing

Ehrman comments that "one of the problems with ancient Greek texts (which would include all the earliest Christian writings, including those of the New Testament) is that when they were copied, no marks of punctuation were used, no distinction was made between lowercase and uppercase letters, and, even more strange to modern readers, no spaces were used to separate words." Ehrman gives a clever example of words in English like

Godisnowhere

that "could mean different things to a theist (*God is now here*) and to an atheist (*God is nowhere*)."

In the case of the New Testament, the preponderance of the number of manuscripts (5,366 in all) that have been found to be 99.5 percent identical attests to the fact that errors did occur. But the fact that 5,366 manuscripts may not be 100 percent identical is overshadowed by thousands of manuscripts to check, compared to hundreds or less for other written history. Ehrman tempers his assertions by stating that "Most of these differences are completely immaterial and insignificant."

Issues of Interpretation

Let's look at Biblical interpretation in the context of Biblical scripture. The first example involves Christ's parable of the mustard seed. Jesus makes the comment that the smallest seed is the mustard seed. See Matthew 13:31-32 and Mark 4:31:

> **Matt 13:31-32:** *The kingdom of heaven is like a mustard seed, which a man took and planted in his field. Though it is the smallest of all your seeds, yet when it grows, it is the largest of garden plants and becomes a tree, so that the birds of the air come and perch in its branches.*

Today it is known that the smallest seed is *not* the mustard seed; the seed of the epiphytic orchid is smaller. That seed is only 35 millionths of an ounce, and it is dispersed through the air like a minute dust particle. Existing in many varieties, it grows on trunks and branches of trees in damp, warm rainforests in Central and South America and in South Africa, not in semi-arid regions like Israel.

From the fact that the seeds of an epiphytic orchid are smaller than mustard seeds, do we conclude that the New Testament is not all true or that the story does not bear meaning? I think not; we have to look at the real point of the story in the context in which it is given. According to the cultural and scientific realities of the time, the mustard seed was smallest. My assumption is that the non-essential matters of faith, used as parables, allegories, metaphors, or rounding of numbers are not Biblical errors but are cultural in nature and directed toward the spirit of the moment. Jesus surely knew that the point of his mustard-seed story would be lost if the name of the epiphytic seed was used, because nobody he was speaking to had heard of the seed, an orchid, or a rainforest. This parable is a metaphor, and what is physically true is not as important as the implied message that the Kingdom of heaven was very small then, similar to a mustard seed—or, in today's jargon, the seed of an epiphytic orchid—but would grow exponentially large in time. We witness that growth today.

It is entirely possible that in the twenty-eighth century, scientists may discover a seed even smaller than the epiphytic orchid seed. Let's use our imagination and call it a *lubac* seed. If we then rewrite Matthew 13:31-32 using the lubac seed instead of a mustard seed, do we change Christ's message in the parable? I think not. Who in the culture or environment of Israel at the time, or in the present, would understand the message?

Let's consider another example of interpretation involving the number of women who discovered the empty tomb:

- In Matthew 28:1, Mary Magdalene and the other Mary discover the empty tomb (two women).
- In Mark 16:1, Mary Magdalene, Salome, and Mary, the mother of Jesus, discover the empty tomb (three women).

- In Luke 24:1, "the women" discover the empty tomb (no specific number is named; let's say *x*).
- In John 20:1, Mary Magdalene discovers the empty tomb (one woman is specified).

From this information, do we conclude that the Bible is in error? I again think not. If four people were assigned to report a news event, I suspect there could be discrepancies in the resulting story. (This morning I read two vastly different descriptions of a win by the Indianapolis Colts.) Four writers describe the event of discovering the empty tomb, mentioning two women, three women, *x* women, and one woman. The issue to me is that an empty tomb was discovered by some number of women, and Jesus was *not* in the tomb. Just because *x* women were listed does not mean that there may have been "several." Such is the nature of Biblical interpretation. To some extent, we should cut God some slack in our interpretation of such an event.

Conclusion

We who speak English are reading one Bible of which there are more than 3,000 English translations and paraphrases. Fallible humans have made these, and humans can make mistakes of copying, translation, and interpretation. *Inherent* means that a trait cannot be considered separately from its nature because of being innate, permanent, or inseparable. Brian McLaren, in his book *The Last Word and the Word After That,* makes I call a faith axiom when he writes, "I want to affirm that my regard for scripture is higher than ever. I would prefer to use the word *inherency* to describe my view of Scripture: God's *inerrant* Word is *inherent* in the Bible." I choose to resolve much of the controversy of inerrancy by adhering to the following assumption, or faith axiom, and moving on. The axiom is supported further in Chapter 4, where we use mathematics to provide evidence of the reliability of the Bible:

> *Faith axiom.* The character of God is not simply the Bible, but the Bible is the most extraordinary way to God's character.

It is my desire to pursue the character of God. Although I consider the Bible the best road, I also consider great sermons, great theology books, walks in beautiful scenery, the birth of a baby, the joy of music, prayer, great conversations with friends, and contemplation to be other ways to God's word. I gain peace about the worth of the Bible by embracing a certain mystery about its content. Such mystery addresses God's infiniteness.

The New Testament gives a clearer revelation of God than the Old Testament. Through the NT scriptures, we learn how to establish an intimate relationship with God. We should not worship the Bible. We should worship God, whose message we receive through the Bible.

Finding Our Own Faith Axioms

In making faith axioms, we wrestle with the dynamic tension between belief and questioning in fields such as science, mathematics, history, and philosophy, and between the points and counterpoints of the Bible. We hope that a resolution of the tension brings us to faith axioms. When we change math axioms, we change the consequences. When we change faith axioms, we change the consequences.

A final comment on the word *proof* or *prove* is in order, especially for mathematicians, before we continue. The apologetic arguments in this book are not deductions of theorems from a finite set of axioms as normally expected in mathematics. Instead, all kinds of arguments—inductive, statistical, and even metaphorical—will be used to point you from a position of nonfaith to a position of faith. The theorems we come to will all be called *faith axioms*. The goal in each chapter on this journey is to provide apologetic evidence for forming faith axioms, thus bringing you closer to that marvelous leap of faith in Christ: We accept what has sufficient evidence to render it probable. It is our goal to shorten the gap between nonfaith and faith as follows:

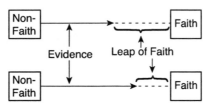

Figure 1.6 From evidence to faith

Let's move now to faith axioms whose apologetic defense rests in a framework of mathematics.

Chapter 2

Paradoxes in Mathematics and Christianity

Paradox is the foundation of all essential truth, and the capacity to embrace paradox is the key to psycho-spiritual growth, whether it be on the golf course, in the boardroom, over the stove, or even in the bedroom.

—M. Scott Peck

A paradox can be philosophical as well as analytical. Although the concept has a base in mathematical logic, the idea of paradox can extend to all aspects of life. Take, for example, Antelope Canyon, South, near Page, Arizona. The canyon is actually a slot in a large sandstone formation, resulting from the erosion of flash floods. It is a great representation of the notion of *paradox*—so narrow, the sunlight of day can barely enter, depending on the time of day and the angle of the sun. A person can walk through this canyon on a winding path, sometimes in light, sometimes in darkness. Similarly, a person moves in life along a spiritual path, winding between darkness and light, believing and questioning. But what a beautiful trip!

The *American Heritage College Dictionary* defines *paradox* as "a seemingly contradictory statement that nonetheless may be true." Many of us associate the words *ambiguity*, *irony*, and *opposing views* with this definition. Paradox as a philosophical concept is related closely to the mathematical concept of contradiction. A contradiction is always false, but a paradox may only seem to be false. Both are central to many arguments in this book. Paradox is a concept that you'll see in every aspect of your life once you recognize it. Let's look at some paradoxes:

- Natural vs. supernatural
- Deism vs. theism
- Believing vs. questioning
- Passing a treadmill test vs. five weeks later having a heart attack

Figure 2.1 Antelope Canyon, South Slot. Photo by Karen E. Bittinger

M. Scott Peck, whom I've deemed the modern-day C. S. Lewis of psychology and religion, says, "Paradox is the foundation of all essential truth, and the capacity to embrace paradox is the key to psycho-spiritual growth, whether it be on the golf course, in the boardroom, over the stove, or even in the bedroom." Life is full of paradox. The capacity to embrace paradox is the key to psycho-spiritual growth.

Embracing Paradox in Daily Life

Let's consider an example of embracing paradox that occurred when I was in graduate school. I took an advanced educational psychology class where the professor taught in an informal style, often using casual conversation. One day, out of the blue, the professor asked me if I had a topic to discuss. He'd never picked out any of us in that manner before, but I was up for a challenge. I went to graduate school with an intense desire to improve mathematics education. Without hesitation, I asked, "How can we get teachers interested in becoming better teachers?" The professor then led a discussion that not only provided an answer to my question but also became a foundation of my teaching, learning, and writing to this day.

First, he asked us to give examples of good teaching. After some discussion, he asked us to give examples of poor teaching. I soon realized that the two discussions formed a paradox. Looking at both the good and bad styles helped us learn the concepts of excellent teaching. The psycho-spiritual growth this discussion provided was a more profound understanding of quality teaching. In effect, the embrace or acceptance of the paradox brought growth to me—I allow myself to maintain the dynamic tension between what seem like contradictory concepts in order to improve my teaching and writing. The resolution comes from its embrace.

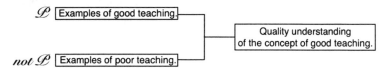

Figure 2.2 $\mathscr{P}$ and *not* $\mathscr{P}$

My favorite number is 8. Why? The answer involves fancies only a mathematician might consider: First, $8 = 2^3$, and the 2 and 3 are different numbers. Second, there are many geometric symmetries in the numeral. It's even the number on my baseball uniform. Now, suppose

I wanted to teach a small child the concept of "the number *eight*." Certainly I would show the child lots of sets of *eight* objects and do some finger counting. But if I were to teach properly, I would also show the child sets of objects whose size was *not eight*. In the process, the child would gain a more profound understanding of the number eight. Only in looking at what the number is and what it is not do we get context, or an understanding of where it fits, better defining what it is. This is another way to look at paradox. When we define what something is and what it is not, then we have a better understanding.

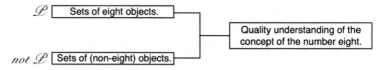

Figure 2.3 Eight and not eight

To look more deeply at paradox in our daily lives, we can examine our professions. Each of us has to deal with satisfying issues as well as difficult issues in our jobs every day. My two favorite questions when meeting a person for the first time are, "What is the most satisfying aspect of your profession?" and "What is the most difficult aspect of your profession?" I try to guess the answers, but inevitably I am wrong on at least one. No one has ever failed to answer *both* questions. And just as we looked at definitions of good and bad teaching, in our own lives, we can look at paradoxical examples from our professions and see them in a new light. By asking these two questions, I think I get to know a person in greater depth, and a meaningful conversation typically ensues. A few months ago, I took a survey of some people I know to see if the theory bears out.

Individual/Profession	𝒜 Easiest/Most Satisfying Part of Profession	not 𝒜 Hardest/Most Difficult Part of Profession
Pete Glogoza, officer, Indiana State Police.	"Being motivated to go to work each day."	"Doing the job, knowing you are making the world a better place. If you don't feel that way, then you shouldn't be in this job!"
Dave Rodriguez, pastor, Grace Community Church, Noblesville, IN.	"The easiest would be weddings and baptisms. The most satisfying are things I do with my gifts and passions, such as preaching and leading, and that which relates to individuals moving closer to God or healing."	"The hardest is dealing with interpersonal conflict and/or helping a family deal with grieving."
Penny Rodriguez, wife of Pastor Dave Rodriguez (above), Grace Community Church, Noblesville, IN. Penny has a profession as a pianist besides that of being a pastor's wife.	"Creating a home environment that is restful for my husband to return to after being in a highly stressful environment. Really, we have been so fortunate to be at GCC…there have been very few bad times as compared to the really good things that have happened. It is so interesting that our son wants to be in the ministry…not many pastor's kids do because of the church environment they have been in!"	"When someone is angry or upset with something we are doing as a church (or at Dave specifically), I tend to take it too personally. I get very defensive and mother-hen-ish."

Individual/Profession	$\mathscr{D}$. Easiest/Most Satisfying Part of Profession	not $\mathscr{D}$. Hardest/Most Difficult Part of Profession
Russ Umphenour, former president, RTM. RTM owned and operated fast-food restaurants.	"Working with people to help them grow."	"Finding good people in a high turnover business."
M. Scott Peck, author of many psycho-spiritual books, among them the *The Road Less Traveled* books.	"Being a successful author."	"Things that go with success. I sometimes feel like a property and that people use and misuse you. Givers come under the guise of taking, and takers come under the guise of giving. People idolize me for all the wrong reasons."
Dr. John H. Isch, retired cardiovascular surgeon, president, CEO, and chairman of the board of directors, CORVASC, Cardiothoracic and Vascular Solutions.	"Playing a significant role in contributing to the correction of an individual's very real and life threatening physical problem. To be able to intervene surgically and mechanically to correct problems that have an immediate and often dramatic result on both clinical symptoms and longevity. Related to this has been the multitude of relationships that have been established from the work environment."	"When patients do not do well as a result of our operative intervention and tragic problems that follow can become life threatening is a very difficult issue to deal with. I have, over time, begun to 'own' the grief and heartache that is subsequently experienced by patients and families going through devastating results. Though enjoying immensely my relationships with a diverse group of people, some of the same people can present the most frustrating, difficult, and challenging aspects of any vocational career."

Al Albert, former TV voice of the NBA's Indiana Pacers.	"When I pull off all the difficult challenges and present a smooth flowing broadcast."	"The element of time and schedule during the season is the most difficult task of my job. I do lots of reading and studying, for example of newspapers off the Internet. The issue is, have I done enough? Then when I do the game, I want to let the game flow and present my knowledge in a timely manner, and not force myself to have to present it all."
Dusty Baker, former manager of the San Francisco Giants and the Chicago Cubs.	"Seeing the progress in players. Seeing them move from young men to mature men to fatherhood."	"Keeping the team playing as a cohesive unit and finding the time necessary to handle all the before and after game activities."
Kathy Ratekin, stay-at-home mother.	"Being there to share each emotion or concern they have through the day. I am there with them when they learn how to take their birth steps and express joy and confidence in them when they learn how to do new things such as ride a bike. I am also grateful that I am with them during the difficult times such as when they get hurt physically or emotionally such as getting picked on by other kids. Nothing can make a child and mom feel better than a great big hug."	"Taking care of kids as infants was very difficult. The stay-at-home mom is always doing everything for them from diapering to feeding while sometimes having very little adult communication with others. I was constantly tired and lacking enough sleep. Thus, I was not full of energy or patience with each of my children."

How do these outstanding individuals embrace paradox? To me, this happens through the commitment they have to excel, serve, and contribute to society. They accept the negative aspects of their work in order to pursue the positives. These people are making positive contributions to society. Some people yield to the negative aspects of a job and either quit or give up on making an impact.

Paradox and Contradiction

People often confuse a paradox (two true statements that seemingly conflict) with a *contradiction*, a statement that expresses opposing statements or viewpoints (one part negates the other). Whereas a paradox can be viewed as a statement that seems contradictory, both parts of a paradox may still be true. Let's look at examples of paradox and then of contradiction to better understand the difference. Some examples of paradoxes that may not really be paradoxes are in the following table:

Paradox	Analysis	Resolution
The Liar paradox: A man states, "I am lying."	If he *is* lying, then he is telling the truth; and if he is telling the truth, then he is lying. If the sentence is true, then it is false. If the sentence is false, then it is true.	This is a contradiction.
Titus 1:12: One of their own men, a prophet from Crete, has said about them, "The people of Crete are all liars...."	If the prophet is telling the truth, then is he a liar as well? If he is a liar, then can you believe his assertion?	This is a contradiction, though it is attributed to the speaker; it is not an example of the Bible being contradictory.
Barber paradox: The barber in Mathville is a man who shaves all and only those men in the village who do not shave themselves. Who shaves the barber?	If he does not shave himself, then he shaves himself. If he does shave himself, then he cannot shave himself. It follows that he both does and does not shave himself.	This is a contradiction. It would have to be reworded to be resolved, say as "The only person in Mathville who shaves himself is the barber; all others are shaved by the barber."
Standing is more tiring than walking.	Walking would seem to require more exertion, but standing may be more tiring because of the element of emotional energy used in waiting, frustration, and so on.	Stand still for one hour. Then, rest and walk for one hour on the same day. On the next day, walk first for one hour, and then stand still for one hour. Compare your tiredness in each instance.

Paradox	Analysis	Resolution
Is that a new antique?	If it is new, is it an antique? If it can be old, how can it be new?	A true antique aficionado would understand the statement.
We will study directly, but first let's sit and talk.	If we first sit and talk, we are not directly studying.	A better sentence would be, "We will study right after we sit and talk." This represents the way paradox pervades everyday life.
Library paradox: City library creates a bibliographic catalog of all (and only those) catalogs that do not list themselves.	Is City library in the catalog? If it is, then is isn't. If it isn't, then it is.	This is a contradiction.
Truth is relative.	If the statement is absolute, then it is relative. If its truth is relative, then it is absolute.	This is a contradiction.

A contradiction is always false. It consists of two statements, one of which negates the other. For example, each of the following is a contradiction:

- $2 + 3 = 5$ *and* $2 + 3 \neq 5$.
- $\sqrt{2}$ is rational *and* $\sqrt{2}$ is irrational.
- The Cincinnati Reds won their baseball game yesterday *and* The Cincinnati Reds lost their baseball game yesterday.
- God created the universe *and* Marv Bittinger created the universe.

In general, every contradiction is a paradox, but not all paradoxes are contradictions; the parts of a paradox just seem to be in opposition.

Connecting Paradox to Mathematics: Proof by Contradiction

How can mathematical truth come out of paradox? One of the most fascinating methods of proof in mathematics is proof by contradiction. In mathematics, a *contradiction* is a statement that is always false; one part of a contradiction negates the other. Thus, a contradiction is an unresolvable paradox. In mathematical logic, a contradiction is a statement of the form

P and (*not P*)

Such a statement is always false, as shown by the following truth table, where a conjunction with *and* is true only when both parts of the statement are true, and the negation *not P* is true when P is false and false when P is true. Note here that the notation for P and *not P* is normal italic, mathematical notation, to differentiate a contradiction from a paradox. The meaning in mathematics is that *not P* is strict mathematical negation of P.

P	*not P*	P and (*not P*)
T	F	F
F	T	F

A *proof by contradiction* of a sentence S is a proof that first assumes *not S* to be true. Then, using any axioms or theorems of the mathematical system considered and rules of mathematical logic, try to find a sentence P such that both

P and (*not P*)

are true at the same time. Intuitively, the sentence S can be only true or false, but not both. If we assume its negation to be true, and this yields another sentence both true and false, then *not S* cannot be true, so S must be true.

Let's consider an example. In the real number system, prove that

S: For every x and every y, if x is rational and y is irrational, then $x + y$ is irrational.

A real number x is rational if and only if it can be symbolized by fractional notation in which the numerator and denominator are integers. That is, $x = \frac{a}{b}$, where a and b are integers and b≠0. These numbers are all rational: $-400, -3, 0, 1, 2, 8, \frac{2}{3}, -\frac{1}{2}, \frac{114}{835}, -\frac{888}{47}, 0.43434343...$, and -0.8625.

The numbers $\sqrt{2}, \pi, \sqrt[3]{5}, -\sqrt{3}$ and $0.383883888388883...$ are all irrational (not rational). (Note here the evidence of learning by embracing paradox as we consider examples and counter-examples of rational numbers.)

Proof. To prove S by contradiction, we first assume *not S* is true. This means that

not S: There exists an x and y such that

x is rational,

y is irrational, and

$x + y$ is *not* irrational (is rational).

Because x and $x + y$ are rational, then

$x = \frac{a}{b}$, for some integers a and b, and

$x + y = \frac{c}{d}$, for some integers c and d

Then

$$(x+y)-x = \frac{c}{d} - \frac{a}{b} = \frac{cb-da}{db}$$

Now the latter expression

$$\frac{cb-da}{db}$$

is also rational because products, differences, and quotients of integers are rational provided divisors are nonzero. But this says that y is rational because $(x + y) - x = y$. We have a contradiction, because we deduced

(y is irrational) and (y is rational)

Thus, *not S* can't be true, so S is true.

Note that a key in doing proofs by contradiction is to first be able to rewrite the negation of the sentence S as we did here. But to develop that topic takes us too far afield from the goals of this book. For more detail, see the bibliography.

An even more paradoxical result can happen when we use proof by contradiction to prove the existence of a mathematical entity without displaying it. Let's consider an example. Let's prove that

> *S:* There exists an irrational number a and an irrational number b such that a^b is rational.

Proof. To prove S by contradiction, we first assume *not S* is true. This means that

> *not S:* For every irrational number a and every irrational number b, a^b is irrational.

The truth of *not S* says that no matter what two irrational numbers we pick, when we raise one to the power of the other, we get an irrational number. We know from real-number theory that $\sqrt{2}$ is irrational. It follows that $\left(\sqrt{2}\right)^{\sqrt{2}}$ is irrational. Then it also follows that

$$\left(\left(\sqrt{2}\right)^{\sqrt{2}}\right)^{\sqrt{2}} \text{ is irrational.}$$

But this also says that

$$\left(\left(\sqrt{2}\right)^{\sqrt{2}}\right)^{\sqrt{2}} = \left(\sqrt{2}\right)^{\sqrt{2}\sqrt{2}} = \left(\sqrt{2}\right)^{2} = 2 \text{ is irrational.}$$

But we know that 2 is rational. The original negation led us to the result that

(2 is irrational) and (2 is rational)

This is a contradiction. So, the negation *not S* is false; and *S*, our original sentence, is true.

The point of fascination is that the proof has shown that there exists an irrational number *a* and an irrational number *b* such that a^b is rational, but it does not produce that number. Actually, it can be proven that $\left(\sqrt{2}\right)^{\sqrt{2}}$ is irrational, but we will not do that here.

A form of paradox occurs to me every time I see or use proof by contradiction in mathematics. It amazes me that to prove *S*, I may first assume *not S* and try to find a sentence *P* such that both *P* and *not P* are true at the same time. You have seen how some truths in life and learning are the result of embracing paradox. Analogously, some truths in mathematics are the result of embracing paradox via proof by contradiction.

The fact that a situation can arise where the existence of something can be proven but not necessarily shown or produced is a thorn in the side of mathematicians. Mathematicians would like to avoid the use of proof by contradiction, but it has so much power that they can't give it up. I assert that by analogy to proof by contradiction, embracing paradox for psycho-spiritual growth is something humans should not give up because of its power.

Paradoxes in Mathematics That Can Be Resolved

A key word in defining a contradiction is *unresolvable*. A *paradox* is a statement that seems contradictory but may still be true. We can think of a paradox in mathematics and everyday life as a possibly resolvable contradiction. Paradoxes occur in everyday life, mathematics, science, religion, philosophy, and especially in Christianity. We will consider some mathematical paradoxes first and then some paradoxes in Christianity. We will try to resolve each paradox either mathematically or philosophically.

The first way a paradox can occur in mathematics is by a mistake in understanding. Let's consider an example.

Paradox 1: Elaine's Paradox

Does $1 = 0.99999\ldots$? It is a proven fact in mathematics that

$$1 = 0.9999999\ldots$$

The paradox:

$\mathscr{P}.\ 1 = 0.9999999\ldots$ *and **not** $\mathscr{P}.\ 1 \neq 0.9999999\ldots$*

My precious wife, Elaine S. Bittinger, is a lovely, modest woman who deserves utter admiration for living with a mathematics textbook author more than 42 years. Elaine has passed this test with flying colors. But she has not studied mathematics beyond high school algebra and geometry. Thus, her mathematical experience is limited. Elaine's paradox was her struggle with the issue of whether $1 = 0.9999999\ldots$. That is, the 9s past the decimal point continue forever. She felt that the sentence was false because when she saw $1 = 0.9999999\ldots$, her mind registered 0.9999999, thinking that the series, or sum, was finite. Her mathematical experience never covered the notion of an infinite series and its limit. Touch on the infinite, and paradox can arise.

Elaine's paradox occurred because of a mistake in understanding. Let's look at some of the arguments I gave Elaine to resolve her paradox.

Argument 1. Elaine, suppose you think that the decimal notation stops at some point, say after six digits. Then

$$1 = 0.999999$$

But $1 \ne 0.999999$ because 0.9999999 is larger than 0.999999 and certainly closer. Elaine dismissed this argument. It simply can't be! There could have been an unwillingness at this point to allow that $0.999999...$ is even symbolism for a number. I was dead in the water with this argument. Let's consider another.

Argument 2. Elaine, do you agree that

$$\frac{1}{3} = 0.333333..., \text{ and}$$

$$\frac{2}{3} = 0.666666...?$$

She agreed with the assertions, drawing on her experiences with long division to think again that the digits stopped repeating. Granted these facts, what happens when we add $\frac{1}{3}$ and $\frac{2}{3}$? She quickly said that we get

$$\frac{1}{3} + \frac{2}{3} = \frac{3}{3} = 1$$

Then, suppose we add the decimals on the right side:

$$\frac{1}{3} = 0.333333...$$

$$\frac{2}{3} = 0.666666...$$

$$\frac{3}{3} = 0.999999... = 1$$

We should still get 1 as an answer. Do you agree? She refused to accept this argument as well.

Argument 3. Eventually, I found a proof that worked. It goes as follows: Elaine, do you agree, using long division, that

$$\frac{1}{3} = 0.333333...?$$

Remarkably, she said, "Yes." Then let's multiply on both sides by 3 and see what we get:

$$3 \cdot \frac{1}{3} = 3(0.333333...)$$

$$1 = 0.999999...$$

Success at last! The fact that I had finally found an argument that Elaine accepted after 42 years of married life *amazed me*. After all those intuitive "no" answers, I finally found one that worked. It was a mathematical and spiritual breakthrough moment!

Paradoxes often occur in probability theory because a person reasons from experience without a mathematical proof.

Paradox 2: The Birthday Problem

Of *n* people in a group, what is the probability that at least two of them have the same birthday (day and month, but not necessarily the same year)? In a room with 30 people, the probability that at least 2 people have the same birthday (excluding year) is 0.706 or 70 percent. The paradox:

> $\mathcal{P}$. In a room with 30 people, the probability that at least two people have the same birthday (excluding year) is 0.706 or 70 percent, and

> *not* $\mathcal{P}$. Because there are 365 possible days for birthdays in a year, the probability of two people having the same birthday is small, something like 30/365~0.082.

The paradox becomes resolvable by delving into probability theory, which is rich in paradoxical problems. Let *E* equal the event of two people in a group of *n* people having the same birthday. The probability of *E* occurring is given by

$$p(E) = 1 - \frac{365}{365} \cdot \frac{364}{365} \cdot \frac{363}{365} \ldots \frac{[365 - (n-1)]}{365}$$

The development of this formula from probability theory is much too lengthy to be considered in this book. The following table and graph show results for groups of size 2 to 100. For 23 or more people, the probability that 2 or more people will have the same birthday is greater than $\frac{1}{2}$ —that is, it is more likely to occur than not!

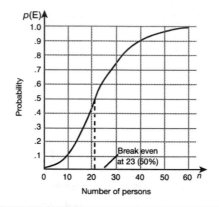

Figure 2.4 Birthday probabilities

n	p(E)	n	p(E)
2	0.00274	28	0.65400
5	0.02710	29	0.68100
10	0.11700	30	0.70600
11	0.14100	31	0.73000
12	0.16700	32	0.75300
13	0.19400	33	0.77500
14	0.22300	34	0.79500
15	0.25300	35	0.81400
16	0.28400	36	0.83200
17	0.31500	37	0.84900
18	0.34700	38	0.86400
19	0.37900	39	0.87800
20	0.41100	40	0.89100
21	0.44400	50	0.97000
22	0.47600	60	0.99410
23	0.50700	70	0.99916
24	0.53800	80	0.999914
25	0.56900	90	0.999994
26	0.59800	100	0.9999997
27	0.62700		

Some paradoxes occur in mathematics through faulty proofs.

Paradox 3: A Faulty "Proof" that 2 = 1

The paradox:

$\mathscr{P}$: $2 \neq 1$ and *not* $\mathscr{P}$: $2 = 1$.

Look over the following "proof." See if you can resolve the paradox.

Proof. Let $a = 1$ and $b = 1$. Then we know that $a = b$, and we proceed as follows:

$$a = b \tag{1}$$

$$ab = b^2 \qquad \text{Multiplying by } b \text{ on both sides} \tag{2}$$

$$ab - a^2 = b^2 - a^2 \quad \text{Subtracting } a^2 \text{ on both sides} \tag{3}$$

$$a(b - a) = (b + a)(b - a) \quad \text{Factoring} \tag{4}$$

$$\frac{a(b-a)}{b-a} = \frac{(b+a)(b-a)}{b-a} \quad \text{Dividing by } (b-a) \text{ on both sides} \tag{5}$$

$$a = b + a \qquad \text{Simplifying} \tag{6}$$

$$1 = 1 + 1 \qquad \text{Substituting 1 for } a \text{ and 1 for } b \tag{7}$$

$$1 = 2$$

The paradox is the false statement $1 = 2$ being true when we know it is false. We find that we really don't have a paradox when we clear up the faulty reasoning. Steps (1) through (4) are mathematically correct. The difficulty occurs with steps (5) and (6), where we divide by $(b - a)$ and simplify. Those with algebra experience continue through the development as they would through the solving of a rational equation. But we are dividing by $(1 - 1)$, or 0. In the theory of the real number system, division by 0 is undefined, or not possible.

You can be deceived into thinking there is a paradox by viewing a faulty drawing. In such cases, seeing is not believing. Many results in geometry and other parts of mathematics start with a drawing. But a drawing is not a proof. Let's consider an example.

Paradox 4: A Geometry Paradox

A square is 8 inches on each side. It is cut along the dotted lines as shown and reassembled to form a triangle. The area of the square is 64, but the area of the triangle is 65. The paradox involves the differences in the areas:

$\mathscr{P}$. The area of the square = the area of the triangle and *not* $\mathscr{P}$. The area of the square ≠ the area of the triangle.

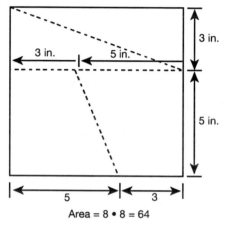

3 in.

3 in.   5 in.

5 in.

5   3

Area = 8 • 8 = 64

Figure 2.5 The area paradox

See if you can resolve the paradox. Try copying the square on a piece of thin paper. Then, cut it out, and rearrange. What happens? Hold a straightedge along the long sides of the supposed triangle. You will note that the lines are not straight. The second figure is actually not a triangle, and thus the formula for the area of a triangle does not apply.

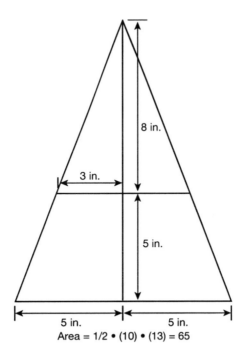

Figure 2.6 Triangle paradox

The American Paradox

You have seen some examples of how paradoxes in mathematics can be resolved. Let's consider some examples in normal life, apart from mathematics.

The following, called The American Paradox, is an excerpt from a letter that's been circulated, shared, and used for years. The source is unknown. It's actually a set of many paradoxes; some of these statements reveal a deep and unfortunate wisdom about the present American society. I have added a few of my own, marked with asterisks (*). In this table, the symbol *not 𝒫* represents a philosophical negation of *𝒫*. Strictly speaking, *𝒫 and not 𝒫* are not strict negations but are reasonably and rationally at odds with each other. Some of these may fall in the realm of emotional arguments, but ultimately they are quantifiable.

The American Paradox

$\mathscr{P}$	and	not $\mathscr{P}$
We spend more.		We have less.
Our wallets are fat.		Our souls are starving.
We buy more.		We enjoy it less.
We have bigger houses.		We have smaller families.
We have more conveniences.		We have less time.
We have more medicine.		We have less wellness.
We read too little and watch TV too much.		We pray too seldom.
We have multiplied our possessions.		We have reduced our values.
These are times of tall men.		Men are short on character.
These are times of steep profits.		These are times of shallow relationships.
Families have two incomes.		Families have more divorce.
We have fancier houses.		We have broken homes.
We have learned how to make a good living.		We have not learned how to make a life.
We added years to life.		We have not added life to our years.
We cleaned up the air.		We have polluted our souls.
We have lots to eat.*		More people in the U. S. suffer from being overweight and obese than any other nation on earth.
We have lots to eat.*		We do not worry about the steady decline of farmers and ranchers and the fact that their death rate is three times the national average.

$\mathcal{P}$	and	not $\mathcal{P}$
We worry about the extinction of the spotted owl and the bald eagle.*		We do not worry about the steady decline of farmers and ranchers and the fact that their death rate is three times the national average.
We outlawed slavery.*		We ignore the plight of migrants and illegal aliens working in meat-packing plants, farms, and sweatshops for practically nothing.
Dog fights and rooster fights are illegal.*		Boxing by humans is legal; participants often suffer brain damage and/or death.
In the World Trade Center disaster of 9/11/01, there were 3,047 deaths.*		In 2000, there were 1,314,000 abortions in the U. S. That's about 3,600 per day.
In the World Trade Center disaster of 9/11/01 there were 3,047 deaths. The nation, with worthy intention, went to military war to stop terrorism.*		Between 1989 and 1998, there were 150,524 deaths from illegal drug abuse. In 2000, there were 601,776 incidences in emergency rooms related to drug abuse. The nation does not go to military war, but the drug infestation of the US continues.
We claim that education is a top priority in our country.*		In many cities, teachers can barely live on their salaries.
Our nation was founded on the motto "In God we Trust."*		We have a separation of church and State, and many lawmakers try to remove God, prayer, and moral concepts from our schools.
Athletes are set on pedestals and accordingly make lots of money.*		The sports pages are 90 percent sports reporting and 10 percent reports of criminal activities by athletes (author's estimate).

You may now be asking how these paradoxes can be resolved, if at all. By *resolved*, I mean by changing one or both parts so both statements are true. To do so, we create a change that takes both truths into account. We did this in each of the four earlier mathematical paradoxes. Resolving these paradoxes might lead to psycho-spiritual growth. Although I make no claims to knowing how to resolve each paradox, here is an all-too-simple resolution of one of them. It would be fascinating to see if it would work. I leave the others to you:

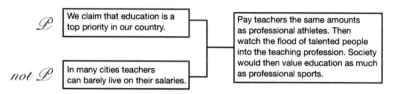

Paradox in Christianity

Being a Christian, I want psycho-spiritual growth. But Christianity is embroiled with paradox. Without delving into the issues in detail, paradox may be the reason people do not choose to become followers of Christ and hold other beliefs. Biblical paradox occurs throughout the Bible. Let's look at some examples of what seem on the surface to be paradoxes in Christianity.

Paradoxes in Christianity

$\mathscr{P}$	*not* $\mathscr{P}$	**Biblical Sources**
God.	Satan.	Hosea 11:9, Gen 1:1, 2 Cor 4:4, Rev 12:9
Wisdom of God.	Wisdom of the world.	Prov 1:7, Psalm 111:10, 1 Cor 1:20, Job 11:6
Right, good.	Wrong, evil.	Rom 12:21, Rom 16:19, Prov 3:7, Amos 5:14

$\mathscr{P}$	not $\mathscr{P}$	Biblical Sources
Energy of the spirit.	Energy of the flesh.	Eph 5:18, Jude 23, Rom 15:19, Eph 2:3
Jesus is fully divine.	Jesus is fully human.	2 Cor 1:19, 2 Peter 1:3, Heb 2:14, 1 John 4:2, Phil 2:6-11, Col 1:15-20
God resides in Heaven.	God resides inside us.	Heb 1:3, Col 1:27
There is one God.	He is three persons.	2 Cor 13:14, John 10:30, 1 Tim 2:5
Humans have free will (human responsibility).	God rules over all creation (divine sovereignty).	Acts 16:31, Joshua 24:15, John 6:44, Rom 9:18
God is perfect in His love.	There is great suffering and injustice in the world.	Rom 8:28, 1 John 4:16, Rom 8:18, 39, 2 Cor 1:7
God acts in the world (He is immanent).	God resides in heaven (He is transcendent).	Matt 6:9
Faith.	Works.	Eph 2:8-10
If you try to keep your life for yourself, you will lose it.	If you give up your life for Christ, you will find true life.	Matt 16:25, Mark 8:35, Matt 10:39, John 12:25, Luke 17:33
The Trinity. God is 3 in 1: the Father, the Son, and the Holy Ghost.	$3 \neq 1$.	Matt 28:19, Luke 3:16, John 16:13-15, Rom 8:9-11, 1 Cor 12:3-6, 1 John 5:6-7 (the word *trinity* does not appear in the Bible; the concept is deduced from scripture)
For when I am weak, then I am strong. (When I become "weak" in the sense of centering my life on Christ, then I become stronger.)	I am only strong when I am strong.	2 Cor 12:10

Let's examine another statement of Peck's and then see how we can embrace paradox to strengthen the Christian faith:

To think paradoxically, we must hold two opposites in our minds simultaneously. I'm not sure we humans are capable of thinking about two things at once and certainly not about opposites. What we can do, however, is bounce back and forth—vibrate—between opposing concepts so rapidly as to make our consideration of them virtually simultaneous. When we learn how to do so, the two will become One and the opposites a Whole.... If you think you have discovered a great truth, and it is not a paradox, then I suspect you may be deceiving yourself.

This quote connects the discussion in Chapter 1 about the road to trust in a faith axiom as the resolution of a dynamic tension between *belief* and *questioning*. Such tension between belief and questioning can lead to the acceptance or resolution of a paradox.

A point of comparison with physics is worth noting here. A *vibration* is a back-and-forth motion or change between two opposing locations in a regularly repeated motion or dynamic resonance. Many examples exist. Among them are the beat of a heart, the vibration of sound, the swing of a pendulum, radio signals, and the change with time of the electric and magnetic fields in an electromagnetic wave. Part of the theory of light involves waves. In each example, we think of the vibration back and forth as one simultaneous entity.

Once, in my high school physics class, we studied the particle theory of light. Light satisfies certain laws as if it's made up of particles, like sand being thrown through the air. We eventually were led to contradictions that put the theory in doubt. Being eager to learn and full of questions, I thought of my own contradiction and asked, "If light is a particle, why doesn't it pile up like grains of sand?" Almost in unison, the class and my teacher laughed at me. Having my questioning spirit snuffed by laughter, especially by the teacher, I never asked questions again. Was that a paradox, and was that why I disliked physics?

Embracing Paradox in Christianity

The Greek mathematician and philosopher Aristotle proposed a way to resolve paradoxes, although he didn't call them paradoxes. He used the idea of extremes. Consider the following extremes, which form a paradox:

Eating nothing vs. eating too much

You can see why this is a little different than a paradox. Instead of "If we eat nothing, we die" and "if we eat too much, we die"—both statements that are provable medically—Aristotle gives us two conditions. Most of us would agree that if we eat nothing, we will starve to death: not a worthy goal. If we eat too much, we become obese and develop heart disease, diabetes, and any number of physical difficulties that also lead to death. A proper balance in our diet resolves the paradox and leads, we hope, to a long life. Aristotle called the proper diet—the resolution—a *golden mean*, and as such does not admit either extreme of not eating or eating too much.

The following table, adapted from Aristotle's book *Nicomachean Ethics,* shows other golden means, or balances, between extremes.

Negative Mean	Golden Extreme	Positive Extreme
Cowardice	**Courage**	**Rashness**
(exhibiting too little confidence)		(exhibiting too much confidence)
Aloofness	**Friendship**	**Flattery**
(showing too little praise)		(showing too much praise)
Self-denial	**Temperance**	**Self-indulgence**
(too little pleasure; no self-indulgence)		(hedonistic; extreme pleasure seeking)
Sloth	**Ambition**	**Greed**
(apathy, inactivity, a disinclination to work)		(excessive desire for wealth or possessions)

Many prominent Christian theologians and ministers resist quick-fix, "how-to" resolutions, such as those of Aristotle, to issues of paradox in the Christian faith. In a 2000 article on Biblical paradox, Richard Hansen proposes three distinct ways for ministers to embrace paradox in their sermons. They can be applied to the Christian faith.

1. Paradox Reframes the Issue

Think carefully about Christ's teaching. Although He presented simple and wise answers in many of His stories, His real solutions were never quick-fix in nature. Being a true Christian is often difficult. Sometimes it's hard to understand exactly what Christ would want us to do in a situation. In many of His teachings, Christ was intentionally paradoxical. Was it because He wanted the listener to make discoveries by investigating, probing, discussing, and struggling with a dynamic tension between belief and questioning?

Christ introduced a classic paradox to His disciples near the end of His time on Earth:

> **Matt 16:25:** *"If you try to keep your life for yourself, you will lose it. But if you give up your life for me, you will find true life."*

> *Paradox.* (If you try to keep your life for yourself, you will lose it) *and* (If you give up your life for Christ, you will find true life).

As we investigate this statement, we *reframe* the paradox by investigating, probing, and discussing.

C. S. Lewis, in his book *Of This and Other Worlds,* says that a *story* may do "what no theorem can quite do. It may not be (like real life) in the superficial sense: but it sets before us an *image* of what reality may well be like in some more central region." We open our imagination to what might be or what might have been.

According to Hansen, "Stories draw us in. We suspend judgment and are more open to change. We move from detached observers to involved participant. The story creates a role for us and we try it on for size." Hansen exemplified his point in the following story in a sermon addressing the previous paradox of faith:

I created a sermon-length story about a woman on a hijacked airplane who must decide whether to identify herself as a Christian when passengers are told all non-Christians are free to leave. Tension builds as the terrorists move toward her seat, forcing each passenger into a bizarre rite of denial by spitting on a picture of Jesus before being allowed to exit to safety.

In her mind, the debate continues—how much action/ effort/commitment does faith demand?—until the hijacker finally arrives to shove the saliva-pocked face of Jesus in front of her and barks, "What about you?"

Quietly, I asked the congregation, "What about me? What about you?" and sat down. Each was forced to confront the cost of faith and add his own ending.

There was no list of how-to steps at the end of the sermon.

When I place myself in this story, I am filled with paradox. On the one hand, I am a Christian and feel I should be proud to assert the fact. On the other side of the paradox, I recall a sermon by Charles Stanley, who asserted that God knows the truth in my heart. Thus, I could say anything to appease the situation. Then I am led to another contradiction: telling a lie, and the guilt of not standing up for Christ. If I lie, I also save myself to further serve God—that's another paradox. In the process, I may also make a fool of the terrorist, whose paradox is fooling himself—ultimately he can never change the faith in my heart, only the words from my mouth. The dynamic tension prevails—we have reframed the paradox.

In a less threatening situation, I believe that the greatest joy I have in life is when I do things for other people with absolutely no hope of getting a word of thanks or praise. In this sense, the "life" I give up is the selfish satisfaction of being thanked or praised. The "life" I gain is being in the will of God or the Kingdom of God in my act of service. This is my reframing of the paradox.

Another example of reframing comes from the movie *Saving Private Ryan*. The character Capt. John H. Miller, portrayed by Tom Hanks, is ordered to risk his own life in the dangerous post-Normandy-invasion European countryside by looking for a young private named James Ryan. War Department policy says that this boy must be found and sent home as the last remaining of four sons killed in action. Captain Miller unquestioningly performs his duty amid the overwhelming desire to survive and return to his wife. Near the end, when Miller is dying, he says to Private Ryan, "Earn this! You can earn it!" Embracing the paradox, that he should die so this one particular boy can live, he challenges Ryan to spend the rest of his life earning what was sacrificed for him. We see this played out at the beginning and the end as Ryan returns to the cemetery at Normandy with his family. The offspring of his family would not have existed were it not for the efforts of Miller and his troops. A reframing or resolution of the paradox turned out to be the kind of life Ryan led thereafter.

Christian theologian G. K. Chesterton has been called the master of irony and paradox. He asserts that "Courage is almost a contradiction in terms. It means a strong desire to live taking the form of a readiness to die." Regarding the "gain life/lose life" paradox, he says, "it is not a piece of mysticism for saints and heroes. It is a piece of everyday advice for sailors and mountaineers." It is also a piece of everyday advice for all Christians!

2. Paradox That Harmonizes

Hansen's second method for embracing paradox is through the notion of the two sides of the paradox working in harmony. He makes an analogy with the vibrations of a tuning fork:

Consider a tuning fork. It delivers a true pitch by two tines vibrating together. Muffle either side, even a little, and the note disappears. Neither tine individually produces the sweet, pure note. Only when both tines vibrate is the correct pitch heard.

Like a tuning fork, harmonious paradoxes declare their truth when two sides of the paradox vibrate in unison. This requires care and honesty. Unlike the tuning fork, which is forged by highly controlled mechanical processes, the paradoxes of Scripture must be forged by the words of highly subjective preachers. Yet despite our biases toward one tine or the other, neither side of the paradox should be muffled, even a little. Paradox beckons us into Mystery, and offers a wholesome reminder that God is infinitely greater than our ideas about God.

The following is an example of a paradox that harmonizes:

Paradox. (God rules over all creation [divine sovereignty]) *and* (humans have free will [human responsibility]).

Hansen expands on this paradox using the conflicts between Job and God (see Job 41:3-6):

At the end of his wrestling with God, Job admits God is unfathomable but also (paradoxically) indicates that he knows God better than he did before.... Regular exposure to paradox challenges Christians early on to exchange faith about God for faith in God, a God who is trustworthy even if often inscrutable.... The two-headed monster on Sesame Street uses exactly this strategy to teach children phonetic pronunciations. One head of the monster says "C..."; the other "...AR." Each head pronounces its syllable with ever-shortening time intervals until the two sounds meld together into a new word: "C...AR," "C...AR," "CAR!"

Another example occurs in physics. Once there were two theories of light: a particle theory and a separate wave theory. Now physicists believe both prevail together—they form a paradox that harmonizes.

3. Paradox That Is Two-Handled

Hansen's third method for embracing paradox is through the notion of a two-handled paradox. A harmonious paradox pushes the two sides together, but a two-handled paradox embraces keeping the two sides apart. We once had a T-shaped garden tool. The handle at the top was connected perpendicularly to a rod that had several prongs at the bottom. Having pressed the prongs through the soil, you took hold of the handle with two hands and twisted the handle, thereby loosening the soil. The right hand pushed in one direction and the left in the other. The longer the handle, the easier it was to apply the torque (force times distance) to rotate it. If you put your hands close to the axis of rotation, the task became more difficult as the two parts of the paradox become closer.

The primary two-handled paradox in the Christian faith is the following:

> *Paradox.* (Jesus is fully divine) *and* (Jesus is fully human).

It is supported Biblically by

> **2 Cor 1:19:** *For Jesus Christ, the Son of God, does not waver between "Yes" and "No." He is the one whom Silas, Timothy, and I preached to you, and as God's ultimate "Yes," he always does what he says.*

Heb 2:14: *"Because God's children are human beings— made of flesh and blood—the Son also became flesh and blood. For only as a human being could he die, and only by dying could he break the power of the devil, who had the power of death."*

In Philip Yancey's book *Reaching for the Invisible God*, he discusses how his struggles with the notions of irony and paradox have been beneficial in building his faith. Yancey quotes G. K. Chesterton:

In a memorable phrase that became the virtual cornerstone of his theology, G. K. Chesterton said, "Christianity got over the difficulty of combining furious opposites by keeping them both and keeping them both furious." Most heresies come from espousing one opposite at the expense of the other.... Uncomfortable with paradox, churches tend to tilt in one direction or the other, usually with disastrous consequences. Read the theologians of the first few centuries as they try to fathom Jesus, the center of our faith, who was somehow fully God and fully man.... Inside every person on earth, we believe [as Christians] the image of God can be found, Yet [paradoxically] inside each person there lives a beast.

The paradox of free will versus God's sovereignty can be considered an example of a two-handled paradox as well as a paradox that harmonizes. (This duality may also be considered a paradox.) It cannot be logically reconciled, much like an electron behaving like a particle and a wave, which is logically impossible. We keep both sides and "keep them both furious":

$\mathscr{P}$. Humans have free will (human responsibility).

not $\mathscr{P}$. God rules over all creation (divine sovereignty).

See Acts 16:31, Joshua 24:15, John 6:44, Rom 9:18

Heresy provides an interesting paradox unto itself. A *heresy* is a doctrine at variance with established religious beliefs. What one religion considers heresy another might consider absolute. And in another situation, it could be exactly opposite. For example, Hinduism has millions of gods, Buddhism has none, and Christianity has exactly one.

In a two-handled paradox, there is no room for the color pink amidst the paradox of red at one pole and white at the other. Such a paradox is a mystery to be savored. I have accepted the following faith axiom as I try to embrace the mystery of God's will: God's will is not a puzzle to be solved. It is a mystery to be savored. I trust God and embrace the mystery.

Think of this axiom as a ride with God on an SUV. In this ride, God and I should expect some bumps in the road, some S-shaped curves, some mountains to climb, and some rivers to ford. But at the end of ride, we will thrill at the goal and look back with understanding.

Paradox upon Paradox

Dayna Curry and Heather Mercer were Christian Missionaries in Afghanistan at the time of the U.S. invasion following the 9/11 disaster. Their goal was to spread the message of Christianity and meet the needs of the poor. The Taliban took them captive after they showed a film called the *Jesus Film* about the life and message of Christ to an Afghan family. United States military helicopters attempted to locate the captives but could not find them. Eventually, Heather found matches in a purse, and they built a bonfire. The searchers saw it and rescued them. Since returning to the U.S., Dayna and Heather have had many speaking engagements and have even written a book, *Prisoners of Hope*. At a recent *Jesus Film* seminar, my wife and I heard Dayna say that she wanted to go back to Afghanistan.

Dayna and Heather were involved in one of the most interesting real-life occurrences of paradox. In their book, we clearly see that on one side, these young women were in a foreign country, Afghanistan, violating the laws of the ruling forces, the Taliban, and those of the Muslim religion. On the other side, they say the following:

We simply were making ourselves available to those Afghans who wanted to know about our faith. Nevertheless, we recognized that if the Taliban perceived us as having broken their law or crossed their line, we would have to be prepared to accept the consequences. In the end, we were willing to take punishment because we really believed God had called us to Afghanistan.... [W]e believe the Afghans—like all people—should at least have the opportunity to hear about the teachings of Christ if they choose.

What is the paradox?

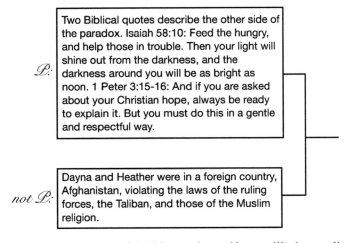

$\mathscr{P}$: Two Biblical quotes describe the other side of the paradox. Isaiah 58:10: Feed the hungry, and help those in trouble. Then your light will shine out from the darkness, and the darkness around you will be as bright as noon. 1 Peter 3:15-16: And if you are asked about your Christian hope, always be ready to explain it. But you must do this in a gentle and respectful way.

$not\ \mathscr{P}$: Dayna and Heather were in a foreign country, Afghanistan, violating the laws of the ruling forces, the Taliban, and those of the Muslim religion.

Let's follow a train of thinking (axioms, if you will) that applies to this paradox, keeping the two sides apart:

A_1 : Dayna and Heather are followers of Christ.

A_2 : They have accepted the Bible as the guide book of their lives.

A_3 : They spread their faith in a gentle and respectful way as described in 1 Peter 3:15-16: "And if you are asked about your Christian hope, always be ready to explain it. But you must do this in a gentle and respectful way." This is the signature quote of Christian apologetics.

A_4 : They were following God's laws, not man's. Acts 4:19-20: "But Peter and John replied, 'Do you think God wants us to obey you rather than him? We cannot stop telling about the wonderful things we have seen and heard.'"

A person standing away from this train of thought might raise questions such as, "The Muslims who destroyed the World Trade Center were following their religion when they broke the laws of the U.S. What's the difference?" Taliban law dictated that Afghan Muslims would be punished by death if they changed religions. Isn't choice a freedom we all should stand for? Isn't learning about another religion a choice we should all have? Why would they want to return to a place where they weren't wanted and were likely to be harmed? Dayna and Heather served without imposing their faith on the people. If the people asked for knowledge, Dayna and Heather shared their faith.

My answer to these challenging questions comes from asking other questions in order to reframe the paradox: Do you want a religion of love and kindness? Do you want a religion that kills people for listening or changing their faith? What are your axioms? I leave you with these questions.

Living in Paradox

One of my mottos is, "I live in paradox!" I may ask to have this placed on my tombstone. There are many examples in my life where God has resolved a paradox, personally, professionally, and spiritually.

My mother died by either suicide or accident when she jumped or fell to her death from a hospital window 10 days after my birth. This tragedy was probably the result of post-partum depression, although such a term was not prevalent at the time. You might construe such an event as the negative side of a paradox. Only in hindsight do I comprehend the loss of my mother. I never knew her. The struggle for me was the lack of a positive, affirming, male role model in the house. My father lived elsewhere, and I was raised by my grandmother. He didn't ignore me, but I yearned to be with him over the years of my youth. My grandfather was in the home where I grew up; but although he was a hardworking provider, he spent virtually no one-on-one time with me. The result was that I grew up seeking approval and looked for it most of all in my teachers, especially my math teachers.

Forty years later, my wife Elaine read in my mother's diary how often she prayed for me. How might God have resolved the paradox of a child losing his mother? Her death resulted in my being raised by my grandmother, a woman who valued a college education. I dare say that had I not been raised in the home of my grandparents on my mother's side, I would not have gone to college. Did the prayers of my mother prompt God to take an ordinary, unpopular high school student and allow him to stumble and succeed in college and in graduate school and then go on to write 192+ math textbooks that are studied all over the world? Ever hear the story about how the turtle got on top of the fence post? Someone had to put him there. God had to put me there.

Here's an example of a more day-to-day experience. Many times, I sense that life is positive (good) and arranged just the way I want it, only to have it rocked by something negative (bad). I often write math books in my mind. I recall one day being lost in my thoughts concerning my writing. A man held a door for me at the YMCA and, being lost in my thoughts, I went right through it without thanking him. He then ripped me verbally. I deserved it! Two sides of the paradox came through: the "good" in my goal of writing well, the "bad" in the kindness I failed to express to a fellow man. Another day-to-day paradox occurs in letters from my students. One tells me how much my books have enhanced her learning. Soon thereafter, I get another, comparing my writing to trash and calling me a "math geek."

In another example, God resolved a paradox by producing the most profoundly spiritual event in my life. In 1998, I had a heart attack. After doctors used stents to open my clogged arteries, I began a very aggressive diet and exercise program to prevent further problems. The "good" was the false sense of security before my heart attack; I had passed a treadmill test with flying colors just five weeks before. The "not good" was the realization that I was indeed mortal, having come close to death.

God resolved the paradox in two ways. First and foremost, I am in better physical health now than ever before in my life. A kind of empowerment comes with exercise. I can do activities like sports and hiking with high endurance. And not only am I more heart-healthy, but exercise also strengthens the body's immune system; inhibits heart disease, diabetes, and dementia; and prevents numerous other health problems.

On a daily basis, God resolves paradox in my life as the ongoing process of my faith journey. I try to pursue my faith through an alternating rhythm of taking hold and letting go, a *believing* and a *questioning*. In this way, I become, as Mark Buchanan describes in Chapter 1, both a hunter and a detective: stalking the evidence, laying ambush, and rummaging for clues as I attempt to resolve paradox on the way to truth.

Chapter 3

Christ: God's Ultimate Resolution of Paradox

If a man wishes to avoid the disturbing effect of paradoxes, the best advice is for him to leave the Christian faith alone.

—Elton Trueblood, *The Yoke of Christ*

We all live amid the paradox of sin. On the one hand, as a Christian, aware of my failings, I feel well qualified to write about sin. But at the same time, writing about sin puts me in the awkward position of judging others. And only God is truly qualified to judge the sins of others.

Humankind lives a paradoxical existence, caught between God-centeredness (goodness) and self-centeredness (sin). The process of becoming a Christian is a turning away from a life of self-centeredness and toward a life of God (Christ)-centeredness. These two concepts form the ultimate paradox of man. You'll see the role Christ plays in resolving this paradox.

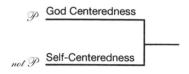

Figure 3.1 The ultimate paradox of man

The meaning of the word *sin* is elusive. In Christian circles, the word describes anything from breaking Biblical and worldly laws to violating commandments. It includes moral wrongdoings to be avoided. Even in less religious circles, the concept of right and wrong extends into ethical and moral dilemmas. The concept of sin has even worked its way into daily vernacular, often when people don't know or understand what it really means. A parent might say "wasting food is a sin" when they really mean that blatantly wasting anything given to us by God is a moral infraction for which we should ask forgiveness.

The word *morals* is disappearing from our society. Even with a resurgent interest in ethics, many schools shy away from teaching morals. The Trustees of the University of Chicago recently issued a proclamation to the effect that teaching morals wasn't a goal of the university. But ironically, Americans proclaim "values" to be important when choosing a President or other political leaders.

Are Sins Just Rules to Be Enforced?

My pastor describes sin as a violation of the character of God. Sin is human rebellion against God and His order for the world. It's an interesting way to tie together God's law—the individual rules we all sometimes worry about breaking—with how those rules violate the overall character of God. When we're self-centered, we're in rebellion against the character of God. When we're God-centered, we're pursuing His character.

In the Old Testament, sin is placed in the construct of the Ten Commandments; they are the safety net—the framework that protects us. God says that if you violate one of these guidelines, you're committing a sin. A violation is a sin against God. You're also creating an unhealthy environment for yourself and the people around you. An act of sin leads not only to your unhappiness, but also to unhappiness for those close to you. The guidelines are meant to protect a society from chaos in much the same way that the laws in our present society keep order. A society with no constructs leads to anarchy—total disorder and confusion. The Ten Commandments can be thought of as God's way of bringing light into a dark place.

Through Moses, God gave the Old Covenant to the Jews. It consisted of many laws or guidelines regarding sacrifices, food preparation, sexual habits, practices of the priests, and so on, as well as the Ten Commandments. Many of the food and sanitation guidelines have been shown by today's medical knowledge to be sound health advice. Much of the rest of the Old Testament was devoted to a long list of Jewish laws (rules) that shouldn't be broken. Keeping God's law maintained an order in society, households, and health. For example, many Kosher laws grew out of basic food-safety issues. But presented as God's law, they were enforced by the Jewish leaders.

Today, modern Christians still use the Ten Commandments as the basis for God's laws to ensure moral solid ground. Although it's obvious that killing, stealing, and adultery are sins, we still try to justify behaviors that we know are morally wrong. Over the years, in a desire to have a finite set of rules to cover all situations, certain churches and groups have extrapolated from the Ten Commandments and used those tenets to pass judgment on their members. My childhood upbringing occurred in such an environment.

The stresses of a conservative religion can take their toll on an immature child. As I look back, my faith was imposed on me by my grandmother, who raised me; my church; a podium-pounding, screaming pastor; and interminable standing pastoral prayers. The notion of sin was a set of rules taught by my grandmother and by my very conservative church. No less important for my grandmother was the notion of Hell. Whether you see Hell as the ultimate lake of fire or as eternal separation from God, for a young boy, it was a threatening deterrent.

Some rules were based on the Ten Commandments. (The "going to Hell" part isn't listed in the Ten Commandments as written in Exodus 20.) For example:

"If you lie, you're going to Hell."

"If you steal, you're going to Hell."

"If you don't respect your parents, you're going to Hell."

Other rules seemed to be part of the culture of my church and upbringing:

"If you smoke, you're going to Hell."

"If you drink alcoholic beverages, you're going to Hell."

"If you don't go to church, you're going to Hell."

My dad didn't smoke during the first 11 years of my life. I can't express the panic and confusion I felt when I walked into a room one day and saw him smoking. My dad was going to Hell, according to the rules of the church!

To become a follower of Christ at my church, you had to "go forward" to the pastor in the front of the congregation, confess your sins, and accept Christ. I did so at the age of 12, but mostly because of the pressure I felt from my grandmother. I wasn't ready, and I didn't understand. It was an agonizing experience. Today, in hindsight, I see the other side of yet another paradox in my Christian life: the love from my grandmother, my father, my pastor, and my friends in the church, even as I was fearful of confessing my sins. Would that love have been there, had I not gone forward? What would have happened if I hadn't responded to the pressure-filled altar call? At the time, I thought my grandmother would stop loving me; but I now realize she would have stayed the course, no matter what. It was another paradox I was too young to resolve.

As we live and grow, we know we will sin. But in the ultimate paradox, we also know Christ embraces us for our faults, for those sins, and forgives us for them. He only asks we acknowledge them and repent, and in exchange, we're rewarded. Again, this is a strange paradox to understand, but it's one we accept with no evidence required in faith.

My rebellion against Christianity as a list of manmade rules to follow began when I was in high school. I was friends with a girl named Rose Wolfe. We talked in study hall or in the cafeteria. I was interested in her in a romantic way, but I was so shy and uncomfortable around girls that I never revealed my interest to her. Knowing she was a member of a Nazarene church nearby, I mustered the courage one Sunday evening to attend her church in hopes of fostering a relationship. I slipped into the back seat of the church right at the start of the service. Little did I know I was walking into the end of a week-long revival. As the guest minister finished his sermon, he began his altar call with pressure-laden pleas that we should go forward and confess our sins. "Did you lust after your neighbor's wife? Did you think about having a drink? Did you smoke even one cigarette?" The guilt trip went on and on.

The people began going forward in a trickle, which increased steadily until people were streaming to the front. The hymn in the background, "Just as I Am," in some psychological way enhanced their guilt trip. The longer the hymn played, the more the preacher rang out his message of guilt, the more people went forward—and for me, the more I resented being called out! I was in an unfamiliar church and not about to confess my sins to these strangers!

But the preacher's victory wasn't to be complete until the whole congregation was in the front of the church—but not me! Finally, the preacher gave up, and I sat there alone in my pew with the entire group—*less me*—standing in front. To top it off, Rose wasn't even there that night! What was the paradox? On one side, I went to the church to pursue this young lady. But on the other side, I got caught in another pressure-laden altar call.

My refusal to go through the motions of another altar call was an aspect of a rebellion that began in my teenage years and continued in college. My buddies and I often made fun of the screaming, overemotive TV evangelists by mocking their pronunciations of *God* as "Gwa-Odd" and *sin* as "See-Yun." Between the ages of 12 and 20, past the time I technically became a follower of Christ, the sermons from the pulpit continued to overwhelm my conscience with the guilt of sin. Another paradox—why does it seem that the longer we wait to ask forgiveness and come back to the fold, the harder it is; and yet Christ accepts us just the same? Yes, when I accepted Christ, my sins were forgiven; but in the days that followed, I inevitably committed more acts of sin.

Somehow there had to be more to God's plan for my life than a continual merry-go-round of sin to guilt to forgiveness, and back around again. Yes, I'm going to sin; and yes, God knows I will; and I'm going to ask for forgiveness, and God knows that too. But there has to be more. *I just know there is more.*

What more is there? As the years have passed, I have come to realize that my passion for Christianity revolves around the following questions and their answers. Why is it that when I wake up each morning, I can't wait to get to my desk to do my writing? Why do I have the inner satisfaction that I chose the right profession when I get a letter or e-mail from a student who has finally found success in math by using one of my textbooks? Why am I so excited about getting back to Utah and Arizona to hike and see my beloved mountains reaching up to God? Why does a piece of music reach in and grab my soul? Why does my heart melt when one of my granddaughters falls asleep on my chest, and I savor the joy of listening to her breathe? Why did my heart skip a beat when I looked across the room the other night in our Bible study and saw Elaine smiling, her beautiful eyes lighting up the room?

One answer to these questions comes from the plan God put in my *heart*. My passion and joy in the faith comes from pursuing God's plan for my life, much more than chasing the sin merry-go-round. Another answer is derived from the so-called love commandments of the New Testament: "Jesus replied, 'You must love the Lord your God with all your heart, all your soul, and all your mind.' This is the first and greatest commandment. A second is equally important: 'Love your neighbor as yourself.' The entire law and all the demands of the prophets are based on these two commandments." (Matthew 22:37-40). Follow God's plans in my heart, and carry out His love commandments—these are the two positive, uplifting goals of Christianity!

The Sin of Self-Centeredness

I'm convinced that sin is an utter reality. I break rules every day—we all do. I speak too sharply to someone. I don't apologize. I don't take that extra minute to listen. I miss an opportunity to share my faith. But in my present, more mature years, I also identify with the definition of sin as self-centeredness. To say it another way, it is a "false self" centered apart from God. Sin can be as overt as plotted acts of murder, rape, theft, child abuse, infidelity, and addiction to drugs and alcohol. A person who commits any of these evil acts is being extremely self-centered, oblivious to what they do to other people.

But sin can also be subtle. It can build up over time until it dominates our lives and leads us down a whirlpool of self-centeredness away from having our minds where God wants them to be. Often, such misdirection is easier to touch or pursue than an abstract, invisible God. In mathematical terms, instead of pointing toward the $(n+1)$st dimension of God's plan, we settle for the nth dimension of self. What are some examples of sins of self-centeredness? Being obsessed with looks, or sex, or conquest can be a form of self-centeredness. Those highly focused on professions and the power they bring over God and family are often choosing money or power over God. And all kinds of recreations can take your eyes off God and His plans, whether it's an obsession over baseball or golf or health or over-indulging children in sports. As D. D. Williams says in his book *Interpreting Theology,*

The core of sin is our making ourselves the center of life, rather than accepting the holy God as the center. Lack of trust, self-love, pride, these are three ways in which Christians have expressed the real meaning of sin. But what sin does is to make the struggle with evil meaningless. When we refuse to hold our freedom in trust and reverence for God's will, there is nothing which can make the risk of life worth the pain of it.

I once knew a lay pastor who told us how he became addicted to money and power. At one time, he thought that if he earned a million dollars, he would have enough money. When he earned the million, he thought he might feel more secure if he earned two million. And so it went; the quest continued. How much is enough? And when that is achieved, what do you do with it? I once heard a humorous, but not necessarily true, story about wealthy oil magnate J. Paul Getty. He was being interviewed about his wealth and was asked "How much money is enough?" His answer was, "Just a little bit more!" Similarly, how much power is enough, how did you get it, and how did you treat people on the way? A God-centered life can give the answers. What's your motive?

Like anyone else, I struggle with self-centeredness. I have a deep and abiding love of baseball. I can watch a game on TV or walk onto a diamond and feel better—the green grass, the groomed infield, the chalk lines, and the crack of the bat. As a youngster, the kids in my neighborhood played ball until we dropped, using taped balls, nailed bats, and ragged mitts. As much as I loved to eat, I was disappointed when called home for a meal. I still watch baseball, and I play the game in my 60s when I attend the Los Angeles Dodgers Adult Baseball camp in Vero Beach, Florida each year. I'm a poor baseball player, but I love to hit and run the bases. (It does happen once in awhile!)

As a child, I wanted to become a major-league baseball player, until I figured out I had no depth perception. (Hitting a curve ball is impossible for me. When a right-handed pitcher releases the ball, it looks like a fastball straight at my head.) I carried out my baseball compulsion as a child by listening to and sometimes attending Cleveland Indians games. In adulthood, the Cincinnati Reds were my favorite. I recall the early years of my marriage, listening to so many Reds games that I neglected Elaine and my family. But in yet another paradox, my sons loved going to Reds games, so I was able to have quality time with them.

When I became more settled in my career, I formed softball teams. As often as three times per week, I participated in softball leagues, again away from Elaine. Even when I got home, I was on the phone, analyzing the games with my players. I think I was filling a need for relationships or approval. Now, I'm not saying that the diversion of playing in a softball league isn't worthwhile or, indeed, a pleasure in God's eyes. But did I need to play in three leagues and expend extensive amounts of time on such a hobby? The need for these relationships and the approval that came from winning were disproportionate. When my obsession became so self-centered and time consuming that it detracted from my marriage, it became a sin.

My obsessive nature wasn't limited to baseball. During the winter, I moved on to bowling. Again, I didn't bowl in a league once a week; it was three leagues per week, together with the purchase of numerous bowling balls, endless lessons in quest of becoming a professional bowler, and trips to bowling tournaments. I became a member of the Professional Bowler's Association, but how much did that quest for approval detract from my marriage and family and from being my authentic self? Being me and believing I was lovable was never enough. All this time, God was asking why I didn't channel some of this passion and energy to Him. Was it a paradox that I'd been given sports abilities but shouldn't use them? Or was it that I needed to focus on God and not self?

The following chart provides a comparison of self-centeredness versus God-centeredness. The targets of self-centeredness can be called *false idols*. M. Scott Peck calls false idols "cheap substitutes for God," and C. S. Lewis calls them "false infinites." The two columns are the sides of a paradox.

not 𝒫. **Man**	*𝒫.* **God**
False idols, cheap substitutes for God. An idol can be anything revered more than God.	Seeking God's plan and will in our lives. (It can, admittedly, be difficult to know God's plan.)
Money for its own sake	Money that may arrive as a result of God's plan first
Fame	Fame that may arrive as a result of God's plan first
Power	Power that may arrive as a result of God's plan first

not 𝒫. Man	*𝒫.* God
Success	Success that may arrive as a result of God's plan first
Poor self-control	Self-control
Self-idolization	Idolization of God only
Arrogance	Self-sacrifice and service for others
Self-interest or preservation	Striving for the approval of God
Striving for the approval of others	
Addictions to drugs, alcohol, materialism, power, sexual conquest outside of marriage	
Doing good things for the wrong motives	
Fantasies	
Perfection	

Let's think of the sin versus God issue in terms of arrows. With a life of self-centeredness, the arrow of your yearnings goes one way and tends to draw you away from God. With a life of God-centeredness, the arrow returns; for me, God is saying that the way I'm living is fulfilling His plan for my life.

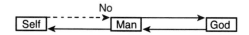

Figure 3.2 Self /Man/God

Why Jesus Christ?

As I recently studied the Old Testament (OT), a pattern seemed to emerge in the way God dealt with His people and the world. This is *my* interpretation of what happened. Keep in mind that I have no way to know the mind of God—this is just my speculation or conjecture: a theory, if you will.

God was loving, and God was just. In story after story, the same pattern emerged. Because I see mathematics in everything, I modeled the stories with the following graph:

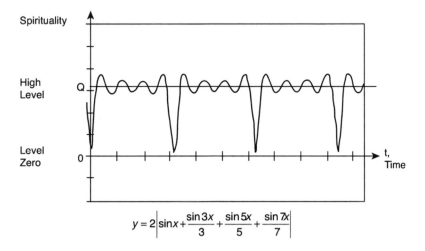

$$y = 2\left|\sin x + \frac{\sin 3x}{3} + \frac{\sin 5x}{5} + \frac{\sin 7x}{7}\right|$$

For those who are interested, the mathematical formula for the model is the graph of the equation

$$y = 2\left|\sin x + \frac{\sin 3x}{3} + \frac{\sin 5x}{5} + \frac{\sin 7x}{7}\right|$$ over the interval from 0 to $\frac{7\pi}{2}$.

Remember, this is just a model and as such isn't the result of research on the part of the author.

Biblical people rejoiced in following God's laws or guidelines, and things went along well for a time; they tended around the high spirituality level Q in the graph. Eventually, the people fell into a pattern of worshiping false idols. Their spirituality fell to a zero level. See, for example, the stories of Othniel, Ehud, Barak, Gideon, Deborah, and Samson in the book of Judges, and Solomon in 1 Kings 11. As I mentioned earlier, Peck calls false idols "cheap substitutes for God" and Lewis calls them "false infinites," considering the real infinite to be God. I call them things that I tell myself "if I can own this" or "if I can do that," I will be more lovable or deserve love, instead of believing that I'm lovable because God loves me just as I am. For the people of the OT, the cheap substitutes for God were gold or wooden statues.

Young in the history of mankind, their way of dealing with an invisible God was to create a statue that seemed easier to see and touch.

Today, our cheap substitutes for God are evidenced by possessions or obsessions. We live in an extreme state of self-centeredness in money, materialism, spite, greed, envy, unfairness, power, sexual conquest, and conceit, all summarized by one word, if you will: *sin.*

When the OT people rejected God's laws and fell into the worship of false idols, God punished them with a flood (as in Noah's time) or the invasion of an opposing nation that destroyed their cities and most of the population, as in the case of the capture of the Jews by Babylon. Soon, God redeemed them with a new ruler, usually military, who emerged as a resolution of the paradox of *man's ways* versus *God's ways,* and the cycle would repeat itself.

God promised Abraham around 2000 BC that the Jewish people would emerge from his patriarchy, endure, and multiply the earth to the end of time. The people made offerings to God on virtually a daily basis to atone for, or "cover up," their sins. The offerings or sacrifices were meant to make amends for the wrongs of the people and reconcile them with God. The sacrifices were often grain and/or the death and burning of animals, such as the prize pick of a herd of cattle or the finest lamb in the lot. Huge was the amount of animal blood spilled at the Tabernacle and/or Temple.

I can't read God's mind, but for me it was almost as if He reached the point of saying, "I've had enough! The Old way isn't working—we need another way; we need a sacrifice that is fully human and fully God." That sacrifice was Jesus Christ. The correspondence between the sacrifices of lambs in the OT with Christ being crucified is evidenced by several references in the New Testament (NT) to Christ as the "Lamb of God." (See John 1:29, 36; 1 Peter 1:19, and Revelation 5:14.) These correspond in the OT to the crossing mark of lamb's blood on the tops and sides of the doors during the Passover in Egypt. (See Exodus 12.) In effect, God replaced the Ten Commandments with a relationship written on our hearts, not on tablets of stone.

From creation, God endowed mankind with free will. God created man to have fellowship with Him. Some ask, why He gave mankind free will—why did He not just have them all love Him like robots? *Paradoxically*, God meant love to come from choice, not by mandate, even if it meant the occurrence of pain and suffering.

Because of man's stubborn free will, he chose to pursue cheap substitutes for God, and self-centeredness, and to move himself away from fellowship with God. This self-will, characterized by an attitude of active rebellion or passive indifference, was evidence of what the Bible calls *sin*. This created the chasm, or paradox, shown here.

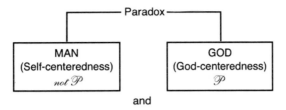

Figure 3.3 The chasm or paradox of man

A gulf ensued between man and God. The sins of man—past, present, and future—eventually had to be paid for with a sacrifice. That sacrifice was the life of God's Son, Jesus Christ. The sacrifice was achieved by one of the cruelest forms of torture and murder ever invented by mankind, Roman crucifixion. In Romans 5:8, Paul writes, "But God showed His great love for us by sending Christ to die for us while we were still sinners." This was God's way of redeeming the paradox, through the life of Christ. Theologian Philip Yancey asserts that "God previewed in His own son the ultimate triumph of His ironic style of redemption." We were on the way to God's resolution of the paradox in which man found himself between self-centeredness and God-centeredness.

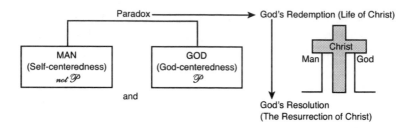

Figure 3.4 God's resolution of paradox

How was the paradox finally resolved? By the resurrection of Christ on the third day—that is, Christ rose from death to life:

> **1 Cor 15:3-6**: *...Christ died for our sins.... He was buried and raised from the dead on the third day ... as the Scriptures said (Isaiah 53:5-9).... He was seen by Peter and then by the twelve apostles (disciples). After that He was seen by more than five hundred of His followers.....*

If we think of this conceptually, it's as if in resolving the paradox, God brought about a turning-around of our self-centeredness toward the nature we should pursue—a God-centered one.

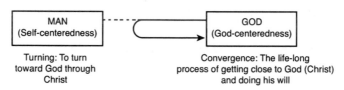

Figure 3.5 Turning and convergence

The act of turning around begins a life-long task of getting close to God and doing His will. I'm doing God's will when I believe I'm loved and let God express His love and relationships through me. Ideally, we get closer and closer, but of course we never make it completely.

Mathematically, if you know some calculus, this convergence can be described as a limit:

$$\lim L(t) = \text{God, where } L(t) = \text{Life's Spiritual Experiences}$$

Although you now understand how Christ and His grace are resolutions to the paradox of sin, consider a slightly different interpretation of Christ as God's ultimate resolution of paradox. I'll use myself (Marv) as an example. (This is a story problem, if you will.) Let's assume that when God made Marv, He manufactured him with an authentic self. Let's go so far as to know that Marv's authentic self is his dedication to writing mathematics textbooks for students so that they can be successful in their studies and go on to worthwhile careers in which they use their authentic, God-given talents to serve humankind. We'll add that Marv's authentic self also consists of a desire to love and nurture his family and friends.

Along the way, Marv, in his humanness, gets distracted, feels unloved, and pursues false idols of withdrawal from relationships, fear of disapproval, not being sure of himself, eating too much food, getting too busy coaching softball teams, writing baseball books, becoming a professional bowler, taking golf lessons, buying things, and so on. But, God wants to restore Marv to his authentic self. How can He do it? One way is to send Christ to renew, restore, redeem, rebuild, and rebirth the authentic Marv who was always there and got lost. In the process, consider Galatians 2:20: "I myself no longer live, but Christ lives in me. So I live my life in this earthly body by trusting in the Son of God, who loved me and gave himself for me."

We can interpret the scripture as follows: I myself no longer live (*Marv's false self leaves*), but Christ lives in me (*Marv gets back to his authentic self and knows that he is always loved by God*). So I live my life in this earthly body by trusting in the Son of God, who loved me and gave Himself for me.

Thereafter, in his humanness, Marv messes up, and he sometimes returns to his false self. How is that dealt with? It is covered by *grace*, the unmerited gift from God, Christ!

The Faith Equation

Man has a choice—he can keep moving deeper into a life of self-centeredness, or he can choose to turn from a life of self-centeredness and move toward the perfection modeled by Christ. The choice is a decision of the *mind*, the *heart*, and the *will* that follows the faith equation:

$$\text{Faith} = (\text{Mind}) + (\text{Heart}) + (\text{Will})$$

According to Hebrews 11:1, "*Faith* is the confident assurance that what we hope for is going to happen. It is the evidence of things we cannot yet see."

According to C. S. Lewis, "The central Christian belief is that Christ's death has somehow put us right with God and given us a fresh start." After coming in faith to Christ and our fresh start, we begin a life-long process of getting close to God and doing His will. A mathematician might call this strengthening of spiritual muscles *convergence*. Theologians call it *sanctification*.

Faith ties in closely to the issue of pain and suffering. According to Yancey,

---

The mature faith ... reassembles all the events of life around trust in a loving God. When good things happen, I accept them as gifts from God, worthy of thanksgiving. When bad things happen, I do not take them as necessarily sent by God—I see evidence in the Bible to the contrary—and I find in them no reason to divorce God. Rather, I trust God can use even those bad things for my benefit. That, at least, is the goal toward which I strive.

Yancey goes on to assert what I accept as a faith axiom: The first principle of the universe is that God is good and loving; anything contradicting that premise must have another explanation.

The events of 9/11/01 provide a prime example. Some believe that all things happen according to God's plan. However, I believe that just because a group of evil, demented terrorists conspired in their free will to blow up two buildings and kill some 3,000 people doesn't imply that God planned the event. Although God may have foreseen what could happen, I can't believe He planned it. As a Christian, I also believe that God took that event and used it to make many people stop and take stock of their spiritual life. Understanding the value of their fellow man can be a positive resolution from this terrible event. Another paradox? Yes, but I wait to see the good offset the pain to the loved ones of those who died. It may be a long time coming, or it may arrive in Heaven. I live with the mystery of this paradox, agreeing with Yancey that "I find no justification for blaming God for what He so clearly opposes.... How can I praise God for the good without censuring Him for the bad? I can do so only by establishing an attitude of *trust* ... based on what I have learned in relationship with God."

I once heard testimony from a religious couple who lost their son in a car wreck; it tested my ability to believe that good can come from tragedy. Their son was married, and his wife pulled up to the accident in her car just after it happened. Their testimony was that they were eventually able to thank God for letting this happen because of the spiritual effects on their lives. *I could never take this viewpoint*! I can see that God might take the tragedy of the paradox and resolve it for the couple by strengthening their faith and their dedication to do God's work. But never, ever, could I thank God for the effect on my life of the loss of a child, no matter how strong it might be. This paradox would go with me to Heaven. When I mentioned this story to my pastor, he thought the couple's reaction was less about acceptance and understanding and more about coping.

Some might say that I'm hypocritical in that I've found over time a way to not only accept my heart attack but also thank God for it. Although it changed my lifestyle, it also unleashed in my heart a passion to use my talents to learn, study, and write from a spiritual standpoint, sharing how I integrate mathematics with my faith. Without the heart attack, the spiritual connection it brought to my wife and me would not have occurred. My deepening relationship with my sons and other male friends would not have happened. I came to realize that I'm always loved by God, no matter what I do. I don't need the approval of man: I have the approval of God. I learned about resolving the paradox of "If I lose my life, I will find my life." I can only be honest and say, *"Thank you, God, for my heart attack!"*

The Paradox of Works vs. Grace

Let's consider another paradox. It involves the issue of *works* versus *grace*:

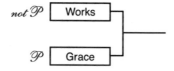

Figure 3.6 The paradox of works vs. grace

In the first 57 years of my spiritual life, I felt caught in this paradox; but in truth, I wasn't aware of my dilemma. How could I get to Heaven? I thought I had to earn Heaven by doing works such as acts of kindness, service, or financial donation. This system of survival was taught by my grandmother. But the trap was determining how many good works were enough. In fact, there were never enough.

To resolve this paradox, I first had to become aware of the notion of *grace*. How often did I hear this buzzword of the Christian faith in my youth, but without any meaningful explanation?

Ironically, it was through a suggestion from my youngest son, Chris, that I read a life-changing book: *The Grace Awakening* by Charles R. Swindoll. I discovered that grace is the *unmerited favor* of God and is the basis of our freedom from sin's captivity, entry to Heaven, faith, and spiritual gifts.

Let's attempt to understand the notion of grace from a Biblical standpoint. One particular Bible quote is the key:

> **Eph 2:8-10:** *God saved you by his special favor when you believed. And you can't take credit for this; it is a gift from God. Salvation is not a reward for the good things we have done, so none of us can boast about it. For we are God's masterpiece. He has created us anew in Christ Jesus, so that we can do the good things he planned for us long ago.*

Swindoll says, "Grace is what God does for mankind, which we don't deserve, which we cannot earn, and which we will NEVER be able to repay." Grace is the free gift we receive from God when we place our faith in Christ—that is, when we choose to accept it. The following are two helpful acronyms, one from Swindoll and the other from Steve McVey in his book *Grace Rules*:

GRACE: *G*od's *R*iches *A*t *C*hrist's *E*xpense

GRACE: *G*iving and *R*eceiving *A*ffirmative *C*hristian *E*ncouragement

Let's develop the concept of grace. One way to understand grace is to look around. For one week, look for situations in your daily life where people perform acts of grace: acts done without any thought of reward. For example, someone might open a door for you without your asking, or let you into a stream of traffic, or return your lost credit card. You'll notice that it becomes much easier for you to go the extra mile as well. Someone might turn in a cell phone or wallet you lost while getting dressed at your health club. One day, at the end of a meal at our favorite seafood restaurant, the waiter refused my credit card, telling my wife and I that an anonymous friend had paid for our dinner. This act of grace occurred two more times when we went back to the restaurant—I started to think all the meals were free. Make a list of acts of grace done for you, and analyze them at the end of the week. Did any act of grace embarrass or overwhelm you? I know the three free dinners overwhelmed us. I think our response came from our feelings of inadequacy; did we deserve those free dinners? I feel this way about God's grace; do we deserve it?

In the second week, make a conscious effort to perform acts of grace for other people, but do them without anyone knowing. For example, pick up a piece of trash in the parking lot, and discard it when no one is watching. Pick up a neighbor's newspaper at the end of the driveway, and drop it at their front door. Philip Yancey and his wife once went through an apartment building occupied by financially deprived senior citizens and put a $100 bill under each door. Oprah Winfrey once told on her TV show of driving up to a toll booth on a Chicago freeway and paying the tolls for the next 10 people behind her. At the end of the week, think back on what you have done, and try to analyze your feelings.

As I look back on my life and my career, I can see many acts of grace—some might call them lucky breaks, but I know they were good things that came to me from people who cared about me and knew the grace of God. They lead me to realize that God's grace is ongoing. He continues to give me gifts that I don't even consider asking Him for. I may not even know they have been given.

Probably the most profound act of grace ever given to me was one I wasn't aware of for 40 years. This act of grace occurred at Ohio State University, where I studied for my master's degree from 1963 to 1965. It takes some telling. My academic paradox began when I graduated from tiny Manchester College, with an enrollment of 1,100 and a caring environment, and entered a huge university like Ohio State, with an enrollment of 50,000 and professors cast in positions of publish or perish. I found good teachers at OSU, but the road was rocky. At the time, the math department was trying to expand its Ph.D. program; toward that end, it accepted 50 graduate students into its master's degree program. Not only was I misplaced in courses with too high of a proof expectation, but the entire group of 50 students was thrown to the wolves with two of the worst professors in the department. The pain was topped off by having no textbook for either course. Of the 50 students, half of us received two Cs or worse in the first quarter and were placed on probation.

In hindsight, I know I should have insisted on being placed in lower-level graduate courses with a more basic proof orientation. To the department's credit, it took the C students and placed them into two recovery courses the next quarter. By God's grace and the teaching of Angelo Margaris, I received an A and a B in each of the two next quarters and pulled myself back to a 3.00 GPA and off probation. (The department also cut my graduate assistantship from $240 per month to $150 per month. I was barely surviving both academically and financially.) I credit Angelo Margaris, for the most part, for salvaging my opportunity to earn a master's degree. His sound teaching of general topology out of a book by George Simmons, complemented with a strong dose of proof techniques, became a mathematics survival handbook for me.

I'm convinced that God has His hand on your life and touches you with grace in ways that you may never know. Two years ago, I tracked down Margaris in Ithaca, New York after his retirement. I thanked him again for his quality teaching and for his support as my master's advisor. Lo and behold, some 40 years later, I found out more about the story of the floundering graduate students. The department's first reaction was to kick all of us out of the program, seeing no hope that any of us could ever get a Ph.D. Margaris said that he and another professor, Jesse Shapiro, went to bat for us and convinced the department to create those two recovery courses that salvaged many of our degrees. To quote Margaris, "Getting a Ph.D. in pure mathematics is so difficult! These people deserved another chance." At least two of us did go on to earn Ph.D.s: myself in mathematics education, and another student in pure mathematics. Margaris touched my life more deeply than I had known. Where would my life be without his act of kindness? This revelation was a touching epiphany.

God's Resolution of the Grace/Works Paradox

The resolution of the grace/works paradox comes by interpreting the notion of *works*, a natural follow-up to turning toward God/Christ as a role model. It forms another faith axiom: I want to do good works because I think it is the right thing to do, not because I want a reward.

A cherry tree produces fruit not because it wants approval, but because it is what God made it to be. Similarly, I want to do good works because that is who God made me to be.

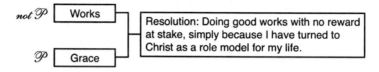

Figure 3.7 Grace and works

In James 2:17, we read, "…it isn't enough just to have faith. Faith that doesn't show itself by good deeds is no faith at all—it is dead and useless." Lewis expands on this in what I call a bias to action: "…if what you call your 'faith' in Christ does not involve taking the slightest notice of what He says, then it is not Faith at all—not faith or trust in Him, but only intellectual acceptance of some theory about Him."

Thus, to me the grace/works issue is a paradox that harmonizes. In the context of the preceding interpretation of works, the two notions merge together in a joyous harmony.

In conclusion, paradox is everywhere in our daily lives and even more so in our Christian lives. Our entire faith depends on it. I agree with Peck and Richard Hansen that paradox should be a subject introduced in grade school or perhaps high school, and then reconsidered in college as contemplation or questioning. Paradox—the idea that two opposing statements can be true—is at the very heart of our relationship with God. It forms a rational basis for our understanding of Christian faith.

Chapter 4

The Probability of Prophecy

> If I am asked, as a purely intellectual question, why I believe in Christianity, I can only answer ... I believe it quite rationally based upon the evidence. But the evidence in my case ... is not really in this or that alleged demonstration; it is an enormous accumulation of small but unanimous facts.
>
> —G. K. Chesterton, *Orthodoxy*

The very popular *Left Behind* series by Tim LaHaye and Jerry B. Jenkins has demonstrated the public's fascination with Biblical prophecy. Although certainly it's the authors' vision of how the Biblical prophecies for the "end times" and second coming of Jesus Christ will happen, it remains to be seen if these prophecies will eventually be filled.

If the *Left Behind* series did nothing else, it caused Christians and non-Christians alike to wonder what the chances are of Biblical prophecy coming true. There are estimates that prophecy occupies from one third to one fourth of the text in the Bible. The problem with being more specific is that not all prophecy can be clearly identified as such. Many of these prophecies have numerous interpretations; others seem more like promises, covenants, or simply predictions. Why is it important as Christians that we examine these passages?

1 Thess 5:20-21: *Do not scoff at prophecies, but test everything that is said. Hold on to what is good.*

Are prophecies directed by the thoughts of men, or is another force at work? Consider this:

> **2 Pet 1:20-21:** *Above all, you must understand that no prophecy in Scripture ever came from the prophets themselves, or because they wanted to prophesy. It was the Holy Spirit who moved the prophets to speak from God.*

Another translation (Holman Christian Standard Bible [HCSB]) says it better:

> **2 Pet 1:20-21:** *First of all, you should know this: no prophecy of Scripture comes from one's own interpretation, for no prophecy was ever borne of human impulse; rather, men carried along by the Holy Spirit spoke from God.*

Reliability of the Bible

To most pastors, the Bible operates as a supreme faith axiom: an overriding means of support. I often yearn to hear them follow a train of thought something like the following. "We should love God with all our heart and soul, and we should love our neighbor like ourselves as described in the New Testament in Matthew 22:37-39. Now, people, this assumes we believe in the reliability of the Bible. In the coming weeks, I'll preach a series of sermons using apologetic arguments to establish the reliability of the Bible."

There are many ways to know the character of God: answers to prayer; the birth of a baby; the effects of relationships; the experiences of life; the beauty of His universe; the joy of music; the sermons of Christian pastors; books by authors like C. S. Lewis, Philip Yancey, Lee Strobel, Dallas Willard; and so on. But the most extraordinary way to know the character of God is the Holy Bible. And I believe in the reliability of the Holy Bible. These are two of my faith axioms.

Countless apologetic books exist to establish the reliability of the Bible. However, as a mathematician, I deal in logic. Logic tells me that if I can encounter evidence establishing that certain points in the Bible are true, then I can reasonably assume that other points in the Bible are true as well—even without evidence. My train of thought goes like this. I believe in the extraordinary reliability of the Bible. One way to confirm the Bible's reliability is to consider the hundreds upon thousands of prophecies that have come true and the likelihood that other, unresolved prophecies will come true as well.

In science, we consider a test *reliable* if it can be given or performed a number of times and yield virtually the same answer each time. I consider the pickups and deliveries of FedEx to be very reliable—that is why I use the company over and over. They come when they say they'll come and deliver packages when they're supposed to. But the $150 inkjet printers I buy are not reliable. Most last less than a year, incur paper jams, and run out of ink too quickly. A repairman is not reliable if he tells me he will arrive at a certain time and more often than not either does not come when he says he will or does not come at all.

How can I verify the reliability of the Bible? One way is to consider the fulfillment of prophecy. A *prophecy* is a prediction of the future, typically a promise made by God through his prophets. If thousands of these promises are fulfilled, it is incredible evidence of the Bible's reliability.

Depending on how you count them, hundreds or thousands of Biblical prophecies have come true. Although many have yet to be resolved—for example, God's promise never to flood the earth after the time of Noah's Ark, and those of the book of Revelation regarding the end times of the earth—we can draw a large data set from the pool of prophecies that have come true.

Suppose it rains today. Is it likely to rain tomorrow? Based on one piece of data, it's hard to come up with a reasonable probability. But suppose it rains every day for 30 days. Then, we might reason that it is likely to rain the next day. This process is called *inductive reasoning*. When we reason inductively, we examine evidence that has occurred, such as the sun coming up each day for thousands of days, and we conclude that it will come up tomorrow. When we reason *deductively,* we use a body of known or assumed facts or axioms and draw further conclusions using logical reasoning. For example, suppose it rained all over Hamilton County in Indiana yesterday. I live in the city of Carmel, Indiana, which is inside Hamilton County. Then, I can deduce that it rained in Carmel yesterday.

Reasoning inductively from the thousands of fulfilled prophecies, and using statistical formulas based on probabilities, I conclude that because prophecies in the Bible have been fulfilled, those prophecies that are not yet resolved will also be fulfilled.

Probability of Prophecy, and an Incredible Paradox

In daily vernacular, the probability of an event E is a number p that is either 0 or 1 or is between 0 and 1: that is, $0 \leq p \leq 1$. The number p represents the chances of an event occurring. But in everyday usage, the notion of chance or odds is subtly different from probability. If the odds of a football team winning a game are 2 to 3, then the probability p of the team winning is

$$p = \frac{2}{2+3} = \frac{2}{5} = 0.40 = 40\%$$

and the probability of the team losing is

$$q = \frac{3}{2+3} = \frac{3}{5} = 0.60 = 60\%$$

Suppose we fill a jar with 70 white jellybeans and 30 red ones. We shake the jar to mix them up, and without looking, we take one out of the jar. We reason that the probability that it is red is 30/100, or 30 percent, and the probability it is white is 70/100, or 70 percent. For an event that cannot happen, the probability is 0. For an event that is sure to happen, the probability is 1. The probability of drawing a green jellybean out of the jar is 0/100 or 0. The probability that it is either red or white is 100/100 or 1.

Are prophecies made by prophets in the Old Testament and fulfilled several hundred years later the result of divine guidance or simply a historical fulfillment related only to chance? No one can prove that these events are not fulfilled by chance, but we can give reasonable probabilities if we wish to assign chance as the moving force in this issue. You can also assign your own probabilities to each event under consideration to provide your own guidance, reach your own results, and make your own conclusions. The arithmetic will seem precise, but the choices are somewhat subjective; you will have to depend on your personal theology and reasoning to make your decision, but we hope it will be a guided choice.

When we reason about the probability of an event, we use *theoretical probability*. If for some reason we can't reason about probability and just repeat an experiment, then we are using *experimental probability*. For the jellybean jar, we reasoned that the theoretical probability of a red jellybean is 30 percent. But in the absence of such thinking, we might repeatedly draw a bean out of the jar, note the color, replace it, and draw another. If we did this 1,000 times and noted that there were 713 red beans, then the experimental probability would be 713/1000 or 71.3 percent. What is the *true* probability? Both kinds of probability are commonly used.

One fascinating aspect of probability occurs when results surprise us. For example, if we found only 17 red jellybeans in 1,000 trials, we would be surprised and perhaps wonder if the experiment had been carried out properly. If all the 1,000 jellybeans we draw were red, we would be really surprised, and it would be a reasonable conclusion that all the beans were red.

It has been shown experimentally that if you kiss someone who has a cold, the probability of your catching the cold is only 0.07, or 7 percent. Have you ever been away from home, met a stranger, started talking about people you know, discovered a common acquaintance, and exclaimed, "It's a small world!" It isn't as small as you think: It has been shown theoretically that the probability of this happening is 0.20, or 20 percent—much more than the virtual 0 you might have guessed.

The preceding are examples of paradox in probability. If we think or guess something might happen with a high probability, but in fact it has a much lower probability (or the converse), then we have a paradox in probability. Remember the birthday problem in Chapter 2? The probability of two people in a group of 30 having the same birthday (year ignored) is 70.6 percent—not, as we might guess, 30/365, or about 8.2 percent.

A *prophecy* is a prediction of the future made by prophets under divine inspiration. Most Biblical prophecy makes a connection between the Old Testament (OT) and the New Testament (NT). For example, when Christ is tempted by Satan, He quotes from OT scriptures. This connects Matthew 4 of the NT with the OT scriptures of Isaiah 11:2 and Deuteronomy 8:3, 6:16, and 6:13.

Other Biblical prophecies make a connection within the NT. This occurs in the Gospels (Matthew, Mark, Luke, and John), linking parts together—especially prophecies by Christ. For example, in Matthew 26:34, Christ predicts that Peter will deny him three times before the cock crows. Later that evening, in Matthew 26:69-75, we find that Peter does deny Christ three times.

Some prophecy is made in the OT about the future but not connected to an event in the Bible. The promise of God after Noah not to flood the earth is such a prophecy. As of now, God has kept this promise, although the entire history of humankind will be required for its evaluation.

Some prophecy is made in the NT about the future. The last book of the Bible is the book of Revelation. Revelation predicts a second coming of Christ; but again, the entire history of humankind will be required for its evaluation.

A Staggering Paradox

In what follows, we step back and examine nine prophecies from the Bible. They are all believed to have been fulfilled, so in that sense all have a probability of 1. We will reason a probability of each just before they were fulfilled, as if we did not know whether they would be fulfilled. Then, we will reason a mathematical probability of all nine occurring, and we will get a number extremely close to 0. Such thinking leads to a staggering paradox, or contradiction. The staggering nature of the paradox underscores our conclusion of the reliability of the Bible.

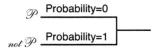

Figure 4.1 A paradox or contradiction: $p = 0$ and $p = 1$

I chose nine prophecies somewhat subjectively, with the main criteria being how they lent themselves to estimating, or reasoning, a probability. I estimate in detail a probability for each of the nine prophecies. For the first prophecy, I contrast my procedure with that of author and astronomer Hugh Ross, founder of an organization called Reasons to Believe, and Henry Morris and Henry Morris III, authors of *Many Infallible Proofs*. I then chose the rest of the prophecies and worked out my own answers using similar techniques.

There is a concept from probability that we use often in these arguments. Suppose an assertion, such as God promising never again to flood the earth after the time of Noah's Ark, occurred t years ago, and to date the prophecy either has not been fulfilled or was just fulfilled. Statisticians would then estimate the probability of the event to be approximately one over twice the number of years: $1/(2t)$. We refer to this as the *time principle* and use it extensively.

We will take some liberties in stating the prophecies in the hope that this will make the process more focused. In so doing, we note that the precise wording of each prophecy may not be found in scripture, but we will provide scripture supporting it.

Prophecy #1: Israel's Messiah Will Be Born in Bethlehem

The Prediction: Mic 5:2: *But you, O Bethlehem Ephrathah, are only a small village in Judah. Yet a ruler of Israel will come from you, one whose origins are from the distant past.*

Ephrathah is an ancient name for Bethlehem and the name of the district containing Bethlehem. This prophecy was written approximately 700 years before the birth of Jesus.

The Fulfillment: Luke 2:4-6: *And because Joseph was a descendant of King David, he had to go to Bethlehem in Judea, David's ancient home.... [W]hile they were there ... [Mary] gave birth to her first child, a son.*

The Probability of Prophecy #1: $\dfrac{1}{10^6}$

Let's consider an argument for the probability of Prophecy #1. We can estimate the probability of the location (where Jesus would be born) and then of the time. Ross and Morris used the following procedures to estimate the probability p_1 of Christ's birth being in Bethlehem:

> **Ross:** If there were 1000 villages in Israel, the rough estimate of the chances of being born in Bethlehem would be 1 in 1000. Thus, the probability would be
>
> $$p_1 = \frac{1}{1000} = \frac{1}{10^3}$$

The key word in this rationale is *If*. I sought to justify the number of villages more precisely.

Morris: Israel at that time was divided into two parts, one redundantly called Israel and the other called Judah. There were two possibilities. Assuming they were equally likely, Morris decided that the probability of the Messiah being born in Judah was 1 out of 2, or $\frac{1}{2}$.

The Messiah could be born either inside Jerusalem or without. Morris again made an assumption of equally likely events and determined that the probability of the birth occurring outside of Jerusalem was 1 out of 2, or $\frac{1}{2}$. One could argue otherwise.

We have the probabilities of each of two events. Assuming the events are independent, the probability of the Messiah being born in Judah and outside of Jerusalem is the product of the probabilities:

$$\frac{1}{2}\cdot\frac{1}{2}=\frac{1}{4}$$

Morris then considered Bethlehem. He asserted that there were "probably" 100 towns and villages in Judah outside of Jerusalem. God might have picked Bethlehem because He knew David originally came from there, and Christ is a direct descendant of David. So, considering the bias, Morris decided instead of 1/100 to use a larger estimate 1/50 as the probability of being born in Bethlehem. Then, assuming independent events, the probability that Christ would be born in Judah, outside Jerusalem, in Bethlehem, is found by multiplying probabilities as follows:

$$p_1=\frac{1}{4}\cdot\frac{1}{50}=\frac{1}{200}\approx\frac{1}{10^{2.3}}$$

Throughout this section, we will focus on the exponent of the power of 10 in the denominator. Taking the common logarithm of 200, we obtain the exponent of 10 needed to express 200 as a power. That is, $\log 200 \approx 2.3$, so $200 \approx 10^{2.3}$.

Ross and Morris derived two possibilities for exponents of 10 in the probability: 3 and 2.3.

I decided to take a deeper look. In the *Barrington Atlas of the Greek and Roman World*, a map of Israel about the time of Christ yields 330 towns or villages. In many translations of the Bible, the phrase "small villages" is replaced by "clans of Judah," which could mean families rather than villages. But if not all clans or villages are listed on the map, then my estimate of 330 towns or villages should be even higher and the probability even smaller. From a population and/or area standpoint, it might be argued that they are not equally likely birthplaces for Jesus. The fact that Bethlehem was so minuscule compared to Jerusalem would make the probability even smaller. Nevertheless, I estimated the probability of Christ being born in Bethlehem to be about 1 out 330:

$$p_1 = \frac{1}{330} \approx \frac{1}{10^{2.5}}$$

This estimate gave me an estimate of about 2.5 for the exponent of 10.

I found another way to estimate this probability in the historical geography *Jews of Palestine*, by Michael Avi-Yonah. He asserts that "In Palestine, west of the Jordan we know of over 2000 ruins of the Roman period, whereas all the place-names quoted in ancient sources do not exceed 500; three-quarters of the places inhabited in antiquity have left no mark in ancient literature." Thus, we might argue that the probability p_1 of Bethlehem being the place of Christ's birth is such that

$$\frac{1}{2000} \leq p_1 \leq \frac{1}{500}, \text{ or } \frac{1}{10^{3.3}} \leq p_1 \leq \frac{1}{10^{2.7}}$$

Comparing all the exponents of 10 we have found—3, 2.3, 2.5, 2.7, and 3.3—I felt confident estimating the probability p_2 of the location to be about

$$\frac{1}{10^3}$$

This prophecy was written approximately 700 years before the birth of Jesus, so the time principle yields another factor p_2 of the probability:

$$p_2 = \frac{1}{2 \cdot 700} = \frac{1}{1400}$$

To find the probability that Christ was born in Bethlehem 700 years after the prediction, we calculate the product p:

$$p = p_1 \cdot p_2 = \frac{1}{10^3} \cdot \frac{1}{1400} = \frac{1}{1000 \cdot 1400} = \frac{1}{10^{6.16}}$$

We estimate $p = \dfrac{1}{10^6}$ to be the probability of Prophecy #1.

It is important to indicate that we are establishing a *cover* estimate for probabilities. By that, I mean that most of the equal signs in the arguments would be better understood and accepted if they were replaced by less-than-or-equal signs: $\leq$. Thus, we have derived a maximum estimate for the probability. That is,

$$p \leq \frac{1}{10^6}$$

which means the probability is at most $\dfrac{1}{10^6}$. It could be smaller.

Surprise: There's a paradox here, or at least an opposing viewpoint. An opposing view of Prophecy #1 regards whether Christ was indeed the Messiah. Those opposing this prophecy might be Jews who do not believe Christ was the Messiah. One reason for Jews' denial was their expectation that the Messiah would come as a ruler on a throne and destroy the Romans. Was Christ the ruler of Israel? To the Jews, Christ was not a ruler in a political sense; to Christians, He was a ruler in a spiritual sense. Some might also question whether Micah was really talking about Jesus of Nazareth.

Prophecy #2: Time of the Birth and Ministry of Christ

The Prediction: Dan 9:25: *Now listen and understand! Seven sets of seven plus sixty-two sets of seven will pass from the time the command is given to rebuild Jerusalem until the Anointed One comes.*

The translation from the older Revised Standard Version (RSV) provides clarity:

Dan 9:25: *Know therefore and understand that from the going forth of the word to restore and build Jerusalem to the coming of an anointed one, a prince, there shall be seven weeks. Then for sixty-two weeks it shall be built again with squares and moat, but in a troubled time. And after the sixty-two weeks ...*

Persia's King Artaxerxes issued the decree regarding the restoration of Jerusalem to the Hebrew Priest Ezra in 458 BC. Then, 483 years later, Christ began His ministry in Galilee. To our point, it is after the restoration of Jerusalem.

The Fulfillment: Matt 2: The date is not explicitly listed there. It must be deduced from the Bible and other historical references.

The Probability of Prophecy #2: $\dfrac{1}{10^5}$

Let's first look at the dates:

$7 \cdot 7 + 62 \cdot 7 = 69 \cdot 7 = 483;$
$-458 + 483 - 1 = 24 \text{ AD};\quad 24 + 6 = 30$
$\qquad\qquad\quad \Uparrow \qquad\qquad\qquad\qquad \Uparrow$
$\qquad\quad$ Date of Ministry$\qquad$ Christ's Age

The age of Christ when he started his ministry was 30. It is generally accepted by historians that Christ was born about 6/5 BC. One year is dropped because the time from 1 BC to 1 AD is just one year, not two.

Let's consider three scenarios and estimate a probability for each:

Time: For Prophecy #2, 458 BC + 2,000 − 1 is 2,457 years, or about 2,500 years. It could have been longer, but to use the time principle, we double 2,500 to get $t=2500$ and $2t=5000$. Then, we use the time principle to estimate a probability for the time that passed until the prophecy was fulfilled:

$$p_1 = \frac{1}{2 \cdot 2500} = \frac{1}{5000}$$

Manner of Execution: The Messiah was God in human form. We estimate the probability of a human being killed or murdered to be about 3 out of 10:

$$p_2 = \frac{3}{10}$$

In fact, the estimate may be much smaller.

When the Destruction Occurred: Relative to the second destruction of Jerusalem, this execution has roughly an even chance of occurring before or after that event: that is, 1 out of 2:

$$p_3 = \frac{1}{2}$$

The probability of Prophecy #2 coming true is (The probability of the time that passed) and (The probability of the manner of execution) and (The probability of the destruction occurring). Assuming independent events, then the probability of Prophecy #2 is estimated by the product

(Probability of A)(Probability B)(Probability C) =

$$p_1 \cdot p_2 \cdot p_3 = \frac{1}{5000} \cdot \frac{3}{10} \cdot \frac{1}{2} = \frac{3}{100,000} = \frac{1}{10^{4.5}} \approx \frac{1}{10^5}$$

Prophecy #3: Zechariah Foretold That Christ Would Be Betrayed for 30 Pieces of Silver

The Prediction: Zech 11:12-13: *And I said to them, "If you like, give me my wages, whatever I am worth; but only if you want to." So they counted out for my wages thirty pieces of silver. And the Lord said to me, "Throw it to the potters"—this magnificent sum at which they valued me! So I took the thirty coins and threw them to the potters in the Temple of the Lord.*

Here, the prophet Zechariah spoke of someone being paid 30 pieces of silver to betray another person. Although the Messiah or Anointed One is not mentioned specifically, this scripture foreshadowed what happened to Christ 500 years later. Zechariah was written approximately between 520 BC and 480 BC. The Messiah lived to 33 AD.

The Fulfillment: Matt 27:3-10: *When Judas, who had betrayed him, realized that Jesus had been condemned to die, he was filled with remorse. So he took the thirty pieces of silver back to the leading priests and other leaders. "I have sinned," he declared, "for I have betrayed an innocent man." "What do we care?" they retorted. "That's your problem." Then Judas threw the money onto the floor of the Temple and went out and hanged himself. The leading priests picked up the money. "We can't put it in the Temple treasury," they said, "because it's against the law to accept money paid for murder." After some discussion they finally decided to buy the potter's field, and they made it into a cemetery for foreigners. That is why the field is still called the Field of Blood. This fulfilled the prophecy of Jeremiah that says, "They took the thirty pieces of silver—the price*

*at which he was valued by the people of Israel—and
purchased the potter's field, as the Lord directed."*

The Probability of Prophecy #3: $\dfrac{1}{10^{10}}$

Zechariah 11:12-13 is generally accepted as a prediction of Christ's betrayal for 30 pieces of silver, about the amount paid for a slave. It turned out to be the same amount paid to Judas when he betrayed Christ. The "potters" mentioned in both scriptures were aliens of low social class. The potter's field was their burial ground. The implication is that Christ was sold out for a sum as small as that used to buy a slave or a cemetery plot for such a slave. The prophet Jeremiah is known for having collected some writings of the other prophets. This may be the reason for his name mentioned in Matthew rather than Zechariah, but admittedly it may be a stretch to make this connection.

Let's consider several scenarios and estimate a probability for each:

> **Time**: Zechariah was written approximately between 520 BC and 480 BC. The Messiah lived to 33 AD. We use 500 years and the time principle to estimate the probability of the event occurring after 500 years to be
>
> $$p_1 = \frac{1}{2 \cdot 500} = \frac{1}{1000}$$

> **Kinds of Payment**: Silver, gold, bronze, land, crops, animals, manual labor—at least seven methods were possible. We estimate the probability of silver for the kind of payment to be
>
> $$p_2 = \frac{1}{7}$$

> **Money Presented**: It was either thrown down or placed. We estimate
>
> $$p_3 = \frac{1}{2}$$

This is most definitely a subjective estimate.

How the Money Was Thrown: The money was thrown down either in the Temple or out of the Temple. We estimate

$$p_4 = \frac{1}{2}$$

Again, this is a subjective estimate.

Amount of Payment: The question is the amount that might cause someone to commit murder. A poor person might murder for a small amount, whereas a wealthy person would not risk being caught unless the payment was very high. Let's say that the payment might range from 1 coin to 1,000 coins, and that Judas was poor. Considering that there were many more poor people than rich, we presume events are not equally likely and estimate the choice of the 30 coins with this probability:

$$p_5 = \frac{1}{100}$$

Betrayal: What is the probability of being betrayed? Admittedly, our choice of a probability here is subjective. We estimate the probability of betrayal as

$$p_6 = \frac{1}{5}$$

Ways of Betrayal: Murder, theft, lying ... the possibilities are almost endless. We estimate the probability to be

$$p_7 = \frac{1}{100}$$

Ways the Priests Might Dispose of the Money: They might have bought business property, farmland, or the potter's field; given it to people outside the priesthood; or bought empty land. They might have kept it or bought some precious possession, such as a robe. Again, the possibilities are almost endless. We estimate the probability to be

$$p_8 = \frac{1}{10}$$

The probability of Prophecy #3 coming true, assuming independent events, is then

$$p_1 \cdot p_2 \cdot p_3 \cdot p_4 \cdot p_5 \cdot p_6 \cdot p_7 \cdot p_8$$

$$= \frac{1}{1000} \cdot \frac{1}{7} \cdot \frac{1}{2} \cdot \frac{1}{2} \cdot \frac{1}{100} \cdot \frac{1}{2} \cdot \frac{1}{100} \cdot \frac{1}{10}$$

$$= \frac{1}{5,600,000,000} \approx \frac{1}{10^{9.748}} \approx \frac{1}{10^{10}}$$

We estimate the combined probability to be about $\frac{1}{10^{10}}$.

Prophecy #4: Manner of Christ's Death

The Prediction: Ps 22:16: *My enemies surround me like a pack of dogs; an evil gang closes in on me. They have pierced my hands and feet.*

Ps 34:20: *For the LORD protects them from harm—not one of their bones will be broken!*

Zech 12:10: *Then I will pour out a spirit of grace and prayer on the family of David and on all the people of Jerusalem. They will look on me whom they have pierced and mourn for him as for an only son. They will grieve bitterly for him as for a firstborn son who has died.*

This prediction was made 500 years before the Roman Empire learned of crucifixion from the Carthaginians. The notion of piercing the hands and feet is just one aspect of death by crucifixion. Psalm 22, in total, contains lots of detail on the crucifixion.

The Fulfillment: Historical knowledge and Biblical quotes:

John 19:31-34: *...they asked Pilate to hasten their deaths by ordering that their legs be broken. Then their bodies could be taken down. So, the soldiers came and broke the legs of the two men crucified with Jesus. But when they came to Jesus, they saw that he was dead already, so they didn't break his legs. One of the soldiers, however, pierced his side with a spear, and blood and water flowed out.*

Luke 23:33: *Finally, they came to a place called The Skull. All three were crucified there—Jesus on the center cross, and the two criminals on either side.*

The Probability of Prophecy #4: $\dfrac{1}{10^{10}}$

Let's consider several scenarios and estimate a probability for each:

Time: Psalms was written between the time of Moses, about 1440 BC, and the Babylonian captivity, 586 BC. We estimate the time passing to be 1,000 years. Then by the time principle, the probability is about

$$p_1 = \frac{1}{2000}$$

Ways to Die: A person could die by natural death or murder. We consider execution by a tribunal to be murder, in a manner similar to the discussion of Prophecy #2:

$$p_2 = \frac{3}{10}$$

Ways of Being Murdered: How could a person be murdered? By hanging, stoning, drowning, spears, bow and arrow, beatings, whippings, being pushed off a high cliff, and so on (see Luke 4:29). When the prophecy was made, the notion of crucifixion had not been invented. This would make an estimate of probability much smaller. We estimate

$$p_3 = \frac{1}{1000}$$

Ways a Body Could Be Mounted for Crucifixion Without Coming Detached: It would feasibly require two nails in the upper body and one or two in the lower body. Possibilities for the upper body are just below the palms of the hands between the small bones of the wrists, arms just above the elbow, and at the shoulders. Note that nails could not be driven through the palms of the hands and still have the body be supported; the nails would tear through the fingers. Possibilities for the lower body are one or two nails in the feet, just above the feet, above the knees, or at the hips. In the case of Christ's crucifixion, one nail was used by placing the left foot over the right and pounding one nail through the arches of the overlapping feet. We estimate $2^9 = 512$ possibilities but allow for the possibility that different combinations hold the body we cut this in half and allow 256 possibilities:

$$p_4 = \frac{1}{256}$$

Not Having to Break Christ's Legs to Be Sure He Was Dead: When a person was crucified, tremendous weight was placed on the upper body—so much so that the body drooped and compressed the lungs. To stay alive, the victim had to push himself up against his legs in order to breathe and not asphyxiate. If the legs were broken, the victim could not push up, and asphyxiation and subsequent death occurred more quickly.

Regarding the scripture "One of the soldiers, however, pierced his side with a spear, and blood and water flowed out," it is common medical knowledge that when a person is dead, there is no blood pressure. Hence, a corpse does not bleed. But in the case of Christ's crucifixion, it is speculated that a blood clot came to His heart because of the scourging, or whip torture, that He incurred before he was placed on the cross. Such a blood clot would cause a gathering of blood and watery fluids in the heart, which would be released by the spear puncturing the heart, much like piercing a balloon full of water.

The assertion in the NT was that the piercing would break no bones. The piercing was to ensure death. To achieve death in one piercing without breaking bones, the spear would just about have to be inserted in the rib and stomach area, which is about 1/8 of the total body area. So, the probability is about

$$p_5 = \frac{1}{8}$$

The probability of Prophecy #4 coming true, assuming independent events, is then

$$p_1 \cdot p_2 \cdot p_3 \cdot p_4 \cdot p_5 = \frac{1}{2000} \cdot \frac{3}{10} \cdot \frac{1}{1000} \cdot \frac{1}{256} \cdot \frac{1}{8} = \frac{3}{4,096,000,000} \approx \frac{1}{10^{10.13}}$$

We estimate the combined probability to be about $\frac{1}{10^{10}}$.

Prophecy #5: Gambling for Christ's Clothing

The Prediction: Ps 22:18: *They divide my clothes among themselves and throw dice for my garments.*

Psalms was written between 1440 BC and 586 BC.

The Fulfillment: John 19:23-24: *When the soldiers had crucified Jesus, they divided his clothes among the four of them. They also took his robe, but it was seamless, woven in one piece from the top. So they said, "Let's not tear it but throw dice to see who gets it."*

This fulfilled the Scripture (Psalm 22:18) that says, "They divided my clothes among themselves and threw dice for my robe."

Mark 15:24: *Then they nailed him to the cross. They gambled for his clothes, throwing dice to decide who would get them.*

The Probability of Prophecy #5: $\dfrac{1}{10^4}$

Let's consider several scenarios and estimate a probability for each:

Time: We estimate the time to be about 1000 BC. Using the time principle, we have

$$p_1 = \frac{1}{2000}$$

Dispersal of the Clothes: One person would take the clothes, or they would be ignored:

$$p_2 = \frac{1}{2}$$

The Method of One Person Getting the Clothes: They might gamble for them, or an officer might assign them to someone:

$$p_3 = \frac{1}{2}$$

The probability of Prophecy #5 coming true, assuming independent events, is then

$$p_1 \cdot p_2 \cdot p_3 = \frac{1}{2000} \cdot \frac{1}{2} \cdot \frac{1}{2} = \frac{1}{8000} \approx \frac{1}{10^{3.9}}$$

We will estimate the combined probability to be about $\dfrac{1}{10^4}$.

Prophecy #6: Manner of Christ's Birth

The Prediction: Isa 7:14: *Look! The virgin will conceive a child! She will give birth to a son, and will call him Immanuel—"God is with us."*

Isaiah was written about 700 BC.

The Fulfillment: Matt 1:22-23: *All of this happened to fulfill the Lord's message through his prophet.*

The Probability of Prophecy #6: $\dfrac{1}{10^{13}}$

The main scenario to consider here is the possibility of one virgin birth. Although it's a physical impossibility but not a spiritual one, we have to remember that per the verse, there was to be only one virgin birth—one out of all the women in the world at the time. The population of the world at the time was about 169,700,000 according to the *World Christian Encyclopedia.* We estimate the number of women to be half that number, 84,850,000, and take the estimate of the probability of a virgin birth to be

$$p_1 = \frac{1}{84,850,000} \approx \frac{1}{10^{7.929}}$$

This estimate is high.

It could be argued that of all the women who ever existed in the world, only one had a virgin birth. From the same source, we consider world populations backward from 2000 AD to 30 AD using increments of at least 100 years, so we can assume people were not counted more than once:

Year AD	Population in billions
30	0.1697
500	0.1920
800	0.2199
1200	0.3619
1500	0.4253
1650	0.5522
1800	0.9026
1900	1.6
2000	6.2
Total	**10.6236**

Considering half of these people to be women, we have a total of 5,311,800,000 women, and we estimate the probability of a virgin birth to be

$$p_1 = \frac{1}{10^{9.725}}$$

This estimate could be even smaller, because it ignores all women who lived before Mary. We can decrease the probability estimate to

$$p_1 - \frac{1}{10^{10}}$$

Time Principle: The book of Isaiah was written about 700 BC. Using the time principle, we get

$$p_2 = \frac{1}{2 \cdot 700}$$

We estimate the combined probability to be about

$$p = p_1 \cdot p_2 = \frac{1}{10^{10}} \cdot \frac{1}{1400} \approx \frac{1}{10^{13}}$$

Prophecy #7: Rebuilding Jerusalem

The Prediction: Jer 31:38-40: *"The time is coming," says the Lord, "when all Jerusalem will be rebuilt for me, from the Tower of Hananel to the Corner Gate. A measuring line will be stretched out over the hill of Gareb and across to Goah. And the entire area—including the graveyard and ash dump in the valley, and all the fields out to the Kidron Valley on the east as far as Horse gate—will be holy to the Lord. The city will never again be captured or destroyed."*

The Fulfillment: Historical. The rebirth became history in 1948 with the founding of the new nation of Israel. The reconstruction of the nine suburbs has gone forward in the locations and in the sequence predicted. The RSV of the Bible refers in this passage to the "brook Kidron." The Kidron Valley lies below the east wall of Jerusalem, separating it from the Mount of Olives on the other side. But it is rare today for a stream or brook to run through the valley because of a tunnel constructed to divert the water.

Jeremiah was written in about 627–586 BC. The guidelines for constructing Jerusalem's nine suburbs were predicted by Jeremiah about 2,600 years ago. He referred to the time of this building project as the "last days"—the time period of Israel's second rebirth as a nation in the land of Palestine, not the end times as predicted in Revelation.

The Probability of Prophecy #7: $\dfrac{1}{10^8}$

Let's consider several scenarios and estimate a probability for each:

> **Time**: The time of the prediction was about 2,600 years ago. The exact date of fulfillment was not in the prediction, but the length of time until fulfillment bears consideration. We estimate that the probability is
>
> $$p_1 = \frac{1}{5200} \approx \frac{1}{10^{3.716}} \approx \frac{1}{10^4}$$

> **Time After Reconstruction**: The time from 1948 to the present is about 50 years, and so far Jerusalem has not been recaptured. Using the time principle, we estimate
>
> $$p_2 = \frac{1}{100} = \frac{1}{10^2}$$

> **Layout of the City**: The area of Israel is about 5,021,510 acres, and the area of Jerusalem is about 30,750 acres. Presuming Jerusalem would be rebuilt where it always stood, the probability that it would be constructed within the parameters listed in Jeremiah 31:38-40 can be estimated by
>
> $$p_3 = \frac{30,750}{5,021,510} \approx \frac{1}{10^{2.213}} \approx \frac{1}{10^2}$$

The probability of Prophecy #7 coming true, assuming independent events, is then

$$p_1 \cdot p_2 \cdot p_3 = \frac{1}{10^4} \cdot \frac{1}{10^2} \cdot \frac{1}{10^2} = \frac{1}{10^8}$$

Prophecy #8: Manner of Christ's Conveyance into Jerusalem

> **The Prediction: Zech 9:9:** *Rejoice greatly, O people of Zion! Shout in triumph, O people of Jerusalem! Look, your king is coming to you. He is righteous and victorious, yet he is humble, riding on a donkey—even on a donkey's colt.*

This prophecy pertains to the manner and means of the conveyance of the triumphal entry of Christ into Jerusalem. It was made about 500 years before this happened.

The Fulfillment: Matt 21:1-11: *As Jesus and the disciples approached Jerusalem, they came to the town of Bethphage on the Mount of Olives. Jesus sent two of them on ahead. "Go into the village over there," he said, "and you will see a donkey tied there, with its colt beside it. Untie them and bring them here. If anyone asks what you are doing, just say, 'The Lord needs them,' and he will immediately send them." This was done to fulfill the prophecy, "Tell the people of Israel, 'Look, your King is coming to you. He is humble, riding on a donkey—even on a donkey's colt.'" The two disciples did as Jesus said. They brought the animals to him and threw their garments over the colt, and he sat on it. Most of the crowd spread their coats on the road ahead of Jesus, and others cut branches from the trees and spread them on the road. He was in the center of the procession, and the crowds all around him were shouting, "Praise God for the Son of David! Bless the one who comes in the name of the Lord! Praise God in highest heaven!" The entire city of Jerusalem was stirred as he entered. "Who is this?" they asked. And the crowds replied, "It's Jesus, the prophet from Nazareth in Galilee."*

See also Mark 11:1-10, Luke 19:28-38, and John 12:12-19.

The Probability of Prophecy #8: $\dfrac{1}{10^5}$

Let's consider several scenarios and estimate a probability:

Time: The prophecy was made 500 years before it occurred. Thus by the time principle,

$$p_1 = \frac{1}{1000}$$

Riding or Not Riding: $p_2 = \frac{1}{2}$

Righteous or Not Righteous: $p_3 = \frac{1}{2}$

Victorious or Not Victorious: $p_4 = \frac{1}{2}$

Ways He Could Ride: Horse, camel, colt, donkey, donkey's colt, carried on a lift, riding in a cart pulled by an animal, and so on. The probability would be at most

$$p_5 = \frac{1}{7}$$

The probability p of Prophecy #8 coming true, assuming independent events, is estimated as

$$p_1 \cdot p_2 \cdot p_3 \cdot p_4 \cdot p_5 = \frac{1}{1000} \cdot \frac{1}{2} \cdot \frac{1}{2} \cdot \frac{1}{2} \cdot \frac{1}{7} = \frac{1}{56,000} \approx \frac{1}{10^{4.748}}$$

We will estimate the combined probability to be about $\frac{1}{10^5}$.

One opposing argument for predicting the conveyance of Jesus into Jerusalem is whether Jesus, who did read and make frequent reference to the Old Testament, chose a donkey to fulfill the prophecy. The Christian assumption is that Christ was God-incarnate. That is, Christ and God were one and the same. If God (Christ) made the prophecy, wouldn't God (Christ) go ahead and fulfill it? The situation to me is like a baseball player hitting a home run over the fence. As he approaches third base, isn't it understood that he will turn and touch home? Have you ever seen a player leave third and head for the dugout?

Lee Strobel quotes noted theologian Louis S. Lapides, who concedes, "For a few prophecies, yes, that's certainly conceivable [that Christ would intentionally fulfill them], but there are many others for which this just wouldn't have been possible. For instance, how would he control the fact that the Sanhedrin offered Judas thirty pieces of silver [Prophecy #3] to betray him? How could he arrange for his ancestry, or the place of his birth [Prophecy #1], or his method of execution, or that soldiers gambled for his clothing, or that his legs remained unbroken on the cross?"

Prophecy #9: Foretelling the Exploits of the Conqueror, Cyrus

The Prediction: Isa 44:28: *"When I say of Cyrus, 'He is my shepherd,' he will certainly do as I say. He will command that Jerusalem be rebuilt and that the Temple be restored."* **Isa 45:1:** *This is what the LORD says to Cyrus, his anointed one, whose right hand he will empower. Before him, mighty kings will be paralyzed with fear. Their fortress gates will be opened, never again to shut against him.* **Isa 45:13:** *"I will raise up Cyrus to fulfill my righteous purpose, and I will guide all his actions. He will restore my city and free my captive people—and not for a reward! I, the LORD Almighty, have spoken!"*

The prophet Isaiah foretold that a conqueror named Cyrus (Isaiah got the correct name) would destroy seemingly impregnable Babylon and subdue Egypt along with most of the rest of the known world. This same man, said Isaiah, would decide to let the Jewish exiles in his territory go free without any payment of ransom (Isaiah 44:28; 45:1; and 45:13). Isaiah made this prophecy 120 years before Cyrus was born, about 170 years before Cyrus performed any of these feats (and he did, eventually, perform them all), and 80 years before the Jews were taken into exile. Isaiah was probably written about 700 BC to 681 BC.

The Fulfillment: Biblical and historical: The first year of Cyrus's reign was 538 BC. In 537 BC, after Persia defeated Babylon, Cyrus the king decreed that Jerusalem and the Temple be rebuilt.

The Probability of Prophecy #9: $\dfrac{1}{10^{15}}$

Let's consider several scenarios and estimate a probability for each:

Major Cities: Let's make a modest assumption of 100 for the number of major cities on Earth at the time. The probability of Babylon being picked as a major city to be conquered would be about

$$p_1 = \frac{1}{100}$$

Conquerors Allowing Captives to Go Home: How many conquerors allow their captives to go home? Not very many. We estimate

$$p_2 = \frac{1}{1000}$$

Cyrus Allowing Jewish Exiles to Return to Israel: How many nations did Cyrus conquer? Among them were the Medes, Hyrcanians, Syrians, Assyrians, Arabians, Cappadocians, Phyrgians, Lydians, Carians, Phoenicians, Babylonians, Bactrians, Indians, Cilicians, Sacians, Paphlagonians, Magadidians, and so on. A reasonable estimate would be 30, because his conquests reached from Persia (Iran) to Egypt. He allowed only the Jewish captives their freedom. We estimate

$$p_3 = \frac{1}{30}$$

Picking the Name: How many names were there be for people in the world at the time? Isaiah gives the name of the ruler, Cyrus—an astounding prophecy that came true! We estimate the probability to be

$$p_4 = \frac{1}{1,000,000}$$

Time: Isaiah made the prophecy about 125 years before Cyrus was born. It was 45 years after his birth that Cyrus performed any of his feats. This was a total of 170 years before the event occurred. By the time principle, we estimate

$$p_5 = \frac{1}{340}$$

The probability p of Prophecy #9 coming true, assuming independent events, is then estimated by

$$p = p_1 \cdot p_2 \cdot p_3 \cdot p_4 \cdot p_5 = \frac{1}{100} \cdot \frac{1}{1000} \cdot \frac{1}{30} \cdot \frac{1}{1,000,000} \cdot \frac{1}{340}$$

$$= \frac{1}{1,020,000,000,000,000} \approx \frac{1}{10^{15}}$$

We estimate the combined probability to be about $p = \dfrac{1}{10^{15}}$.

The Probability of All Nine Prophecies Together

Let's consider the probability of all nine prophecies being fulfilled. Assuming independent events, we multiply all the probabilities (adding the powers), and get

$$\frac{1}{10^6} \cdot \frac{1}{10^5} \cdot \frac{1}{10^{10}} \cdot \frac{1}{10^{10}} \cdot \frac{1}{10^4} \cdot \frac{1}{10^{13}} \cdot \frac{1}{10^8} \cdot \frac{1}{10^5} \cdot \frac{1}{10^{15}} = \frac{1}{10^{76}}$$

How can we understand a number as large as 10^{76}— a quattuorvigintillion—and as small as $\dfrac{1}{10^{76}}$?

Think of a very large space like an enclosed sports arena, and also think of a grain of sand. For example, consider a domed football and sports stadium. Let's try to estimate the number of grains of sand that might fit in such a stadium.

First, we'll take out all the inside objects, such as stands, seats, and locker rooms. Then, we'll fill the space with sand. The approximate dimensions are

$$H = 2316 \text{ in., } L = 9264 \text{ in., } W = 7716 \text{ in.}$$
$$\text{Volume} = 1.65550051 \times 10^{11} \text{ in.}^3$$

A medium grain of sand is about 0.02 inches in diameter. Thus, the volume of a medium grain of sand is about

$$V = \frac{4}{3}\pi\, r^3 = \frac{4}{3} \times \pi \times (0.01)^3 \approx 4.18879 \times 10^{-6} \text{ in.}^3$$

Next, we approximate the number of grains of sand in the stadium by dividing the arena's volume by the volume of a grain of sand:

The number of grains of sand in the stadium

$$= \frac{1.65550051 \times 10^{11}}{4.18879 \times 10^{-6}} \approx 3.952216535 \times 10^{16}$$

We want to express this as a single power of 10. We take log base 10 on both sides:

$$3.952216535 \times 10^{16} = 10^x$$
$$\log\left(3.952216535 \times 10^{16}\right) = x$$
$$17 \approx x$$

Consider $\frac{1}{10^{17}}$. In terms of probability, suppose the stadium is filled with white sand. One grain of red sand is placed in with the white sand and mixed in thoroughly. Then, a person is blindfolded and told to roam through the sand and pick out one grain at random. The probability of a correct pick of the one grain of red sand is $\frac{1}{10^{17}}$.

Recall that the probability of all nine prophecies coming true is $\frac{1}{10^{76}}$. This is much, much smaller than $\frac{1}{10^{17}}$.

Consider that $76 = 4 \cdot 17 + 8$. Then

$$\frac{1}{10^{76}} = \frac{1}{10^{4 \cdot 17 + 8}} < \frac{1}{10^{4 \cdot 17}} = \frac{1}{10^{17}} \cdot \frac{1}{10^{17}} \cdot \frac{1}{10^{17}} \cdot \frac{1}{10^{17}}$$

To get a physical understanding of this large number, think of the probability of selecting the red grain of sand four times. You then get an idea of how impossible it is for all nine of those prophecies to come true.

How does $\frac{1}{10^{76}}$ compare with probabilities of other events involving large-numbered denominators? The probability of winning a state lottery is about $\frac{1}{10^7}$, which is about the same as selecting the red grain of sand out of a fruit jar—a much smaller volume than the stadium. For just our nine considered prophecies, the probability $\frac{1}{10^{76}}$ is like one person winning more than 10 state lotteries. Still pretty unreachable odds, right?

Assuming the age of the earth is 15 billion years (a number subject to debate), about 10^{20} seconds, there is a lot of sand in the stadium. There are 10^{66} atoms in the universe and 10^{80} subatomic particles, and $10^{66} < 10^{66} < 10^{80}$. I am again astounded by the size of the number 10^{76}.

How Many Prophecies Are in the Bible?

We have shown that it is staggeringly improbable for just nine prophecies to come true: $\frac{1}{10^{76}}$. What about considering all the prophecies in the Bible? There are many more than nine, but just how many is subject to debate. Scholars agree that there is no consistent way to count the prophecies, especially if one prophecy is subsumed in another. If we take Ross' estimate that there are about 2,500 prophecies (2,000 of which he believes have been fulfilled), Ralph Muncaster's number of more than 1,000 (with 688 fulfilled), and another data point from Morris of 300, and use a probability of $\frac{1}{10}$ for each of them, we can create a range to work from: $\frac{1}{10^{2000}}$ to $\frac{1}{10^{300}}$. This is even more staggering.

Of all the OT-NT prophecies, we considered just nine and determined that the probability of all nine occurring is about $\frac{1}{10^{76}}$. Because

$$\frac{1}{10^{76}} = 10^{-76} = \underbrace{0.000000\ldots000001}_{76 \;\; decimal \; places, \; 75 \; zeros,}$$

I would call the probability of such an event virtually zero.

How do we account for an event that has probability 0 by the math but probability 1 according to what happened? The answer to this paradox is to say a miracle occurred. The word *miracle* conjures great controversy. Francis Collins, in his book *The Language of God*, says, "… it is crucial; that a healthy skepticism be applied when interpreting potentially miraculous events, lest the integrity and rationality of the religious perspective be brought to question." For me, the thousands of prophecies that have come true, with none yet failing, is sufficient evidence to establish the reliability of the Bible. The evidence is underscored or emphasized to a staggering degree by making the mathematical computations. The proof gets better when we see the paradox. Depending on the faith of the reader, the Biblical results and the mathematics lead me to a leap of faith that some sort of divine intervention occurred in the writing of the Bible. Discussing the controversial topic of *miracle* further is beyond the scope of this book. I am led to make the following faith axioms:

1. The phenomenon of fulfilled prophecy constitutes a unique and powerful evidence of the reliability of the Holy Bible.
2. The remaining unfulfilled prophecies regarding the second coming of the Messiah in the Bible will be fulfilled, in particular those of the Book of Revelation.

We should pay careful attention to what those prophecies predict.

A final opposition to our reasoning might arise in the mind of a reader with a background in the theory of probability and statistics. It concerns the assumption of independent events when obtaining the overall probability of the nine prophecies being fulfilled by multiplying the nine individual probabilities. You might reasonably assume that if one prophecy has been fulfilled, it might increase the likelihood that a second prophecy will be fulfilled; and if two prophecies have been fulfilled, it is even more likely that a third prophecy will be fulfilled. In general, does the fact that some prophecies are known to have been fulfilled increase the probability that subsequent prophecies will be

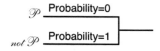

fulfilled? When we consider this possibility using an argument based on conditional probability, the probability is increased from 10^{-76} to 10^{-74}, which is negligible. The details are on the website.

On Seeing Overt Signs from God

We considered nine prophecies that occurred with probability 1. We reasoned mathematical estimates of those same prophecies and found the probability to be 0. The staggering nature of this paradox of results underscores our conclusion of the Bible being reliable. To me, this conclusion is an overt sign from God, although admittedly it is not the same as God standing in front of me and affirming the reliability of the Bible.

Let's discuss the notion of overt signs. Bertrand Russell (1872-1970) was a well-known mathematician, philosopher, and atheist. Al Seckel, in the preface to his book *Bertrand Russell on God and Religion*, writes, … there is a marvelous anecdote from the occasion of Russell's ninetieth birthday that best serves to summarize his attitude toward God and religion. A London lady sat next to him at this party, and over the soup she suggested to him that he was not only the world's most famous atheist but, by this time, very probably the world's oldest atheist:

What will you do, Bertie, if it turns out you're wrong?'
she asked. 'I mean, what if—uh—when the time comes,
you should meet Him? What will you say?' Russell was
delighted with the question. His bright, birdlike eyes
grew even brighter as he contemplated this possible
future dialogue, and then he pointed a finger upward and
cried, *'Why, I should say, God, you gave us insufficient
evidence.'*

It's common to hear comments that suggest God should be more
overt in letting us know He exists. I once read a story that went something like the following. Over a period of six months, a gathering of
stars way out in a distance galaxy began changing into a new formation. At first, only astronomers were aware of the formation. As
it became more pronounced, the event was revealed to the scientific
community. As time passed, the formation became larger and moved
toward the Earth. Eventually, the media alerted the populace, as they
do when a comet travels near our world. The formation came closer and
closer to the Earth, becoming brighter and more prominent until finally
it hovered like an umbrella, creating constant daylight. It spelled out
the following statement:

This message is from God!

I do exist. I implore you to have faith in me!

How would people react? Certainly, Christians and those who
believe in one God would be elated and might celebrate by proclaim-
ing to those who didn't believe in God, "See, we've been telling
you!" Others on the borderline might come to faith. But many non-
believers would look for reasons to dismiss such an overwhelming
sign, especially if the sign in the sky persisted for some time. Those
scientists who are atheists might even try to conjure reasons from
astrophysics why the event happened in order to dismiss it from being
supernatural.

To demand that God gives us a sign as overt as this is to require that He perform on our terms. He chooses to perform on His terms. Although mankind did not choose to have them happen, there have been many Biblical instances where God was overt in His message to mankind. These, of course, depend on your faith in the Bible. In the time of Moses (see the book of Exodus), God turned a river to blood, allowed frogs to infest the Nile river and then the homes of the Egyptians, released gnats and flies into the land, killed the livestock, produced boils on the people and animals, rained down hail across the land, covered the land with locusts, and later replaced light with total darkness for three days. Finally, God killed the firstborn son of each Egyptian family, sparing the sons of the Israelites who followed God's directive to cover their front door with a cross of lamb's blood. To cap all this, God parted the Red Sea, allowing the Israelites to pass through, but closed the sea on the Egyptian army who tried to stop their escape.

As if these signs were not sufficient, later, near Mt. Sinai, God spoke to the people through a cloud, directing them *not* to worship false idols. Would you pay attention to the very overt sign of God speaking through a cloud? I'd like to think I would, but I might decide that my self-centeredness was more important and look for a reason other than God. To me, these were all clear and direct signs of God's existence, as astounding as the star formation story would be if it were true.

Moses then climbed Mt. Sinai in order to receive the tablets with the Ten Commandments. The Israelites were now without their leader, Moses, and for a short time they heard and saw nothing from God. In their frustration at not experiencing God, they decided to collect gold, melt it, and build a golden calf, which they began to worship. God's signs were staring them in their faces, but they found a reason to avoid them. In the New Testament, Christ performed many signs, or miracles" such as bringing dead people back to life, turning water into wine, and feeding thousands of people with a few fishes and loaves of bread. But the most noted, overt sign was his resurrection. The point is, we have seen signs, but we often choose to ignore them. Although we don't see signs today as overt as those in Moses' time, we have to make some effort to find them. Such was the case in this chapter, and such is the case when we study the scriptures and other theological books.

No matter what evidence we encounter, some of mankind will choose to ignore it. For me, the fulfilled prophecies mesh the Old with the New Testament and come about as close to being an overt sign from God as I have had in my lifetime. The fact that the mathematics revealed a paradoxical result underscores the evidence, making the reliability of the Bible a bright, visible sign for me that I can't ignore. To quote well-known apologetic scholar Norman L. Geisler in Strobel's book *The Case for Faith*, "... the only thing we have to look at is what book in the world has been miraculously confirmed. There's only one, and that's the Bible...by the fulfillment of predictive prophecies, and by the miracles performed by those purported to be speaking for God.... Mathematics has shown there's absolutely no way they [the prophecies] could have been fulfilled by mere chance."

Chapter 5

Modeling Growth in Christian Evangelism

Jesus said, "And the Good News about the Kingdom will be preached throughout the world, so that all nations will hear it..." **—Matthew 24:14**

I define an *evangelist* as a messenger who promotes a particular product. Bijoy Goswami, in *The Human Fabric*, describes an *evangelist* as one of three core energies in people and society. In today's world, *evangelical* and *evangelism* are unfortunately terms expressing criticism or disapproval, and they raise red flags. But if I discovered a new ice cream parlor, and I spread the "good news" of its existence, you might be pleased to learn about it—especially if they had a new flavor, *cake batter*. Most people don't understand that a Christian evangelist is someone who delivers a message of love and hope from God, which is not reason for scorn. It is worth noting that Muslims also believe in evangelism; for this reason, conflict between Islam and Christianity is more inevitable than between some other religions.

Everyone who's defended Christian evangelism has heard arguments for and against it. The argument in favor goes like this: "The only way to get to heaven is by following Christ." Many Biblical scriptures support this claim. Two are as follows:

> **John 14:6:** *Jesus told him, "I am the way, the truth, and the life. No one can come to the Father except through me."*

> **1 John 5:11-12:** *And this is what God testified: He has given us eternal life, and this life is in his Son. So whoever has God's Son has life; whoever does not have his Son does NOT have life.*

But often, a person responds, "What about the natives in a tribe in the middle of a jungle? They will never hear the words of Christianity!" But now, we have an apologetic response. The evidence provided by mathematical models herein attests to the fact that remote tribes will indeed hear the message, as will the rest of the world's population. The models attest to the possible fulfillment of the prophecy stated in the epigraph for this chapter:

> **Matt 24:14:** *Jesus said, "And the Good News about the Kingdom will be preached throughout the world, so that all nations will hear it..."*

I approached Professor Jyoti Sarkar, a Christian from India and a research statistician at my university, with ideas for a statistical analysis. We created mathematical models to fit data regarding world population growth (W), Christian population growth (C), and the evangelized population growth (E). We used these models and other evidence to make a prediction about a possible time when the world might be evangelized. Talk about a prophecy being fulfilled! Not only does the Bible say the word of God will be spread throughout the world, now we can also estimate mathematically when that will occur.

As we know, a Christian is a follower of Christ, according to the books of the New Testament—more specifically Matthew, Mark, Luke, and John. However, any person is evangelized if they have heard the message of Christ, even if they are not necessarily a Christian. All Christians are evangelized persons, but not all evangelized persons are Christians. As Christians, we are directed by the Bible to share our faith with the people of the world. The following Biblical quote is often referred to as "The Great Commission":

Matt 28:19-20: *Therefore, go and make disciples of all the nations, baptizing them into the name of the Father and of the Son and of the Holy Spirit. Teach these new disciples to obey all commands I have given you. And be sure of this: I am with you always, even to the end of the age.*

Evangelism is not part of all religions. For example, most religious Jews consider being Jewish a genetic, inherited rite of passage. The notion of being recruited or evangelized to the Jewish faith is abhorrent at best. Although some people who are not born Jewish choose to follow the Jewish religion, to most Jews you are either born to the faith or you aren't.

But I wondered whether it could be deduced when all people in the world will have heard the Christian message.

Models of Growth from Real-World Data

A mathematical model can help to solve the mystery conveyed in the prophecy of Matthew 24:14. In order to derive a model, we analyzed data regarding world, evangelized, and Christian population growth. The variables are defined as follows.

The *independent* variable is

- t = time, or date, AD

The *dependent* variables are

- $W(t)$ = the world population, in billions, at time t
- $E(t)$ = the evangelized population, in billions, at time t
- $C(t)$ = the Christian population, in billions, at time t

Our goal was to create a model that we could scale up to today's numbers. In order to create this model, we wanted to see what data was available that showed us early Christian population growth. How did the word spread 2,000 years ago? With little accurate or exact data in the book of Acts, I turned to other sources for data on our variables during the relevant time periods.

The *World Christian Encyclopedia* contained more adequate data, as shown in the following table.

Table 5:1   World, Christian, and Evangelized Populations, Data in Billions

Date, *t*	W	C	E
30 AD	0.1697	1.2×10^{-8}	0.0008
100 AD	0.1815	0.001	0.0508
300 AD	0.1920	0.0199	0.0672
500 AD	0.1934	0.434	0.812
800 AD	0.2199	0.0495	0.0682
1000 AD	0.2692	0.0504	0.0673
1200 AD	0.3619	0.0701	0.0941
1350 AD	0.3597	0.0867	0.1007
1500 AD	0.4253	0.0810	0.0893
1650 AD	0.5522	0.1169	0.1365
1750 AD	0.7207	0.1600	0.1860
1800 AD	0.9026	0.2082	0.2458
1850 AD	1.2039	0.3278	0.4573
1900 AD	1.6199	0.5581	0.8317
1970 AD	3.6100	1.2166	2.2181
1975 AD	3.9667	1.3168	2.5737
1980 AD	4.3739	1.4327	2.9933
1985 AD	4.7811	1.5486	3.4459
2000 AD	6.2596	2.0199	5.2208

The Mathematics of the Curve Fitting: Basic Approach

In statistics and algebra, we work with many types of functions that can serve as models. We often refer to their graphs as *curves*. Some of these models are linear, with straight-line graphs; some are quadratic, with curves that are U-shaped; some are exponential, and rise steadily and sharply; some are logistic, and are S-shaped; some are bell-shaped, like a normal distribution in probability; and so on. For details, see *Algebra and Trigonometry: Graphs and Models.*

To generate our growth function, Sarkar first examined the data for World Population (*W*). Examining a scatterplot of the natural logarithm (*ln*) of the data, it appeared that some kind of exponential function might fit the data, but we noted a bump around 1350 AD. This bump was created by high growth and then a drop before the population grew again. This was the time of the Bubonic Plague, which destroyed three-quarters of the population of Europe and parts of Asia. For this reason, we felt we should ignore all data prior to 1350 AD. Given that decision, an exponential function seemed to be a good fit But, further statistical analysis was necessary.

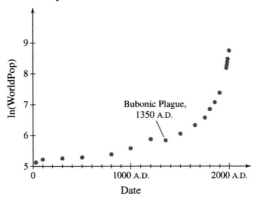

Figure 5.1 Scatterplot of the natural logarithm of the world population data

A quartic polynomial has degree 4, which means that the largest exponent of an x term is 4. The following are examples of quartic polynomials:

$$x^4 - x^3 + 2x^2 - 5x + 7, \text{ and } 0.113x^4 + 2.345x^3 - 8.8x^2 + x - 1$$

We determined that we could fit a quartic polynomial to $\ln(W)$. This is a common procedure for highly accelerated growth patterns in data. Thus,

Fitted World Population function = $\ln(W)$ = a quartic polynomial

Then, using the definition of the natural logarithmic function, it follows that

Fitted World Population = $e^{\text{(Fitted Quartic Fuction)}}$

where e = base of the natural logarithmic function

≈ 2.71828

Before carrying out the regression, we transformed the data in order to make the coefficients of the polynomials have significant digits. We defined the variable t as

t = time, or date AD

The variable x after a linear transformation is given by

$$x = \frac{t - 1750}{100}$$

Carrying out the regression using MINITAB, a statistical software product, we found a model for the Fitted World Population, W:

$$\ln W(x) = 20.4099 + 0.31793x + 0.09319x^2 + 0.034611x^3 + 0.005038x^4$$

The following models for the Fitted Evangelized Population (E) and the Fitted Christian Population (C) were found in a similar manner:

$$\ln E(x) = 19.0956 + 0.55625x + 0.15911x^2 + 0.037903x^3 + 0.005604x^4$$
$$\ln C(x) = 18.9217 + 0.52043x + 0.13126x^2 + 0.018266x^3 + 0.001964x^4$$

Then, using the definition of the natural logarithmic function, it follows that

$$W(x) = e^{\left(20.4099+0.31793\,x+0.09319\,x^2+0.03461\,x^3+0.005038\,x^4\right)}$$

$$E(x) = e^{\left(19.0956+0.55625x+0.1591x^2+0.037903x^3+0.005604x^4\right)}$$

$$C(x) = e^{\left(18.9217+0.52043x+0.13126x^2+0.018266x^3+0.001964x^4\right)}$$

We finally arrive at these growth models:

$$W(t) = e^{\left(20.4099+0.31793\left(\frac{t-1750}{100}\right)+0.09319\left(\frac{t-1750}{100}\right)^2+0.03461\left(\frac{t-1750}{100}\right)^3+0.005038\left(\frac{t-1750}{100}\right)^4\right)}$$

$$E(t) = e^{\left(19.0956+0.55625\left(\frac{t-1750}{100}\right)+0.1591\left(\frac{t-1750}{100}\right)^2+0.037903\left(\frac{t-1750}{100}\right)^3+0.005604\left(\frac{t-1750}{100}\right)^4\right)}$$

$$C(t) = e^{\left(18.9217+0.52043\left(\frac{t-1750}{100}\right)+0.13126\left(\frac{t-1750}{100}\right)^2+0.018266\left(\frac{t-1750}{100}\right)^3+0.001964\left(\frac{t-1750}{100}\right)^4\right)}$$

Depending on your mathematical background, these may be the most complicated exponential functions you have ever seen. We have created models of the three population growths. They help us work with the mystery of the prophecy in Matthew 24:14, because they allow us to pursue finding a date, if such exists, at which the population of the world and the evangelized population will be the same. Do these graphs have any points of intersection? Because of the size of the numbers, it was easier to examine the graphs of the natural logarithms of the functions.

We found that the Christian growth function (*C*)—the Christian population—never intersects the graphs of the world population and the evangelized population, *W* and *E*. That is, according to our mathematical models, the world will never consist entirely of Christians, nor will every evangelized person ever be a Christian. This is confirmed in Biblical writings, especially those in Revelation.

But we made a fascinating discovery regarding the World and Evangelized functions. A point of intersection appeared on the graphs at $x = 2.83$—in actual time, $t = 2033$ AD. That is, according to our mathematical models, the world will be evangelized by the year 2033. This means the world may be evangelized within the lifetime of many people reading this book.

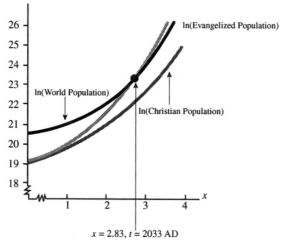

$x = 2.83, t = 2033$ AD

Figure 5.2 Graphs of the natural logarithms of the Fitted World, Christian, and Evangelized Populations

How Quickly Can Evangelism Happen?

Although it's all well and good to have a statistical model that tells us when the world will be evangelized, is it realistic to believe that the word of God can spread quickly enough to work? What is happening today that can give us faith that the entire world population will have heard about Christianity in a relatively short time period—by 2033?

One project working to reach out around the globe is the *JESUS Film Project (JFP)*.

This international endeavor is a project of Campus Crusade for Christ, an organization with many suborganizations, dedicated to evangelism. JFP distributes the film *JESUS*, a two-hour docudrama about the life of Christ, based on the Gospel of Luke. Since its initial release in 1979, the film has been used by the *JESUS* Film Project and more than 1,500 Christian agencies, has been translated into more than 940 languages, and has been seen in *every country* of the world—more than 6 billion viewings worldwide. The goal of the JFP organization is to reach every nation, tribe, and people, in every tongue, helping them see and hear the story of Jesus in a language they can understand. Whether a person speaks Swahili, French, or a language whose name is extremely difficult for most to pronounce, he or she will encounter the life and message of Christ in a language "of the heart." Even the lip-syncing is computerized so it looks as real as possible. As a result, more than 176 million people have indicated their decision to become followers of Christ.

Dedicated workers load movie projectors and screens onto donkeys or vehicles and travel into mountain, jungles, and other remote locations to visit tribes who may never have experienced electricity, let alone a film. They are captivated by what they see. As a result, there has been an incredible rise in the world's evangelized population, to the point that more than two-thirds of those who are evangelized live in the Third World.

A chart that tracks film viewership and a rise in decisions of viewers to become followers of Christ shows a compelling argument for effective evangelism around the world. Many missionary experts have acclaimed the *JESUS* film as one of the greatest evangelistic success stories of all time.

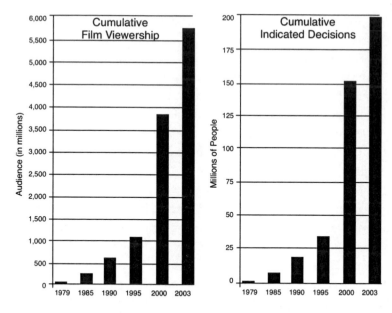

Figure 5.3 Film viewership and indicated decisions

With the incredible advances in technology since 1979, organizers have found they have an even better ability to quickly spread the word. Internet and satellite transmissions have allowed Christian information of all types to be spread throughout the world. In Hawaii, Christian businessmen donated the money to send a *JESUS* film to *every* mailbox in the state. In Morocco, virtually every rooftop has a small satellite dish. In other countries, like China, video compact discs (VCDs) translated to Mandarin and Cantonese are becoming prevalent, as are digital compact discs in other parts of the world. China's Bureau of Church Affairs recently granted permission for *JESUS* VCDs to be reproduced and distributed in *all* provinces of this, the world's most populous nation. The VCDs are embedded in the front and back cover of a hardback book containing the Gospel of Luke in both Mandarin and English. VCDs in Cantonese are also being distributed. Internet technology has also enhanced the visibility of the film throughout the world.

Is Another Prophecy About to Be Fulfilled?

How does all this relate to the prophecy set forth in Matthew 24:14 that all the world will be evangelized? Can we show evidence that this prophecy will come true? Let's look at a prophecy not yet fulfilled:

The Prophecy: The Good News about the Kingdom Will Be Preached Throughout the World (the World Will Be Evangelized)

The Prediction: Luke 21:33: *Jesus said, "Heaven and earth will disappear, but my words will remain forever."* **Matt 5:18***: Jesus said, "I assure you, until heaven and earth disappear, even the smallest detail of God's law will remain until its purpose is achieved."* **Matt 24:14***: Jesus said, "And the Good News about the Kingdom will be preached throughout the world, so that all nations will hear it; and then, finally, the end will come."* **Mark 13:10***: Jesus said, "And the Good News must first be preached to every nation."*

The Fulfillment: The life of Christ was roughly 5 BC to 33 AD. Our mathematical prediction is that this prophecy will be fulfilled in 2033, about 2000 years following the resurrection of Christ. That the prediction is close to being fulfilled is supported by our mathematical models, as well as the JFP: 6 billion viewings, as well as the spread of knowledge through the incredible advance of technology.

The Probability of the Prophecy: Although it isn't our focus in this chapter, if we consider that 2000 years have passed, then by the Time Principle (Chapter 4), the probability that this prophecy will be fulfilled can be estimated to be

$$\frac{1}{4000} \approx \frac{1}{10^{3.6}}$$

We again have the possibility of an extremely unlikely event occurring.

Look again at the Bible quotes (prophecies):

Luke 21:33: *Jesus said, "Heaven and earth will disappear, but my words will remain forever."*

Matt 5:18: *Jesus said, "I assure you, until heaven and earth disappear, even the smallest detail of God's law will remain until its purpose is achieved."*

Jesus spoke these words of prediction in 30–33 AD, saying that His words, His teachings, and His stories will last *forever*. Look again at the table earlier in this chapter. Christianity started in 27–30 AD with very few people in an isolated part of the world. Who would believe then that 2000 years later, these predictions would come so close to being fulfilled?

When I was sharing the preceding result with my youngest son, Chris, then a minister for Campus Crusade for Christ, he said with great joy, "Dad, do you know the significance of that date? It's 2,000 years after the resurrection of Christ!" Most people think Christ was resurrected in 33 AD, but He was actually born in 6–5 BC, so our 2,000-year connection is not quite correct; but it was close enough that I was thrilled with our results. When I went back to tell Sarkar about the resurrection connection, he was like a little kid expressing his joy, saying, "Gee, we got this result, and we didn't fudge the data!"

Not All Models Are Perfect

In the interest of providing the other tine of the paradox, some scientists might come up with opposing views to our analysis. I identify three issues that arise concerning the mathematical models. The first concerns babies born near 2033. Clearly, they will not be evangelized, and the model is flawed from that standpoint. So, in truth, part of the population will not yet be evangelized at that date.

The second is a question of whether a logistic curve might better fit the data. Sarkar eliminated that possibility early on in the research. He used statistical techniques to show that the logistic fit was not appropriate. A logistic curve rises and levels off approaching a boundary line (an asymptote). (We have not included the techniques used for its elimination.) The curves shown earlier in this chapter rise steadily, exponentially if you will, and have no maximum bound.

Third, the model has no relevance beyond 2033. That is, it makes no sense that more people can be evangelized than exist in the world. But by then, the question of whether the world has been evangelized will already have been answered, at least according to the models. Most mathematically derived models like this one have certain numbers that must be restricted from their application.

Another question is the use of the word *evangelized*. The word *evangelist* means a messenger who spreads the "good news," which is the message of Christ's love and hope found in the New Testament books Matthew, Mark, Luke, and John. But how intense has the message been? Was it as simple as seeing the *JESUS* film, or as profound as an intense Bible study that examines the scriptures in great detail and provides extensive time for questioning?

Other questions may arise in the interpretation of the scripture:

Matt 24:14: *And the Good News about the Kingdom will be preached throughout the world, so that all nations will hear it, and then, finally the end will come.*

Does this mean that 100 percent of the world has heard the good news? Or, would an approximation of 99.5 percent be acceptable? Did Christ mean utter precision when He spoke those words, or was He speaking figuratively? Does "really close" suffice? Did He mean the world in those times, or was it the world in future times, which is how we have interpreted it? Or, was the world He was speaking of the Roman empire?

If *evangelized* means *heard the good news*, then what good news have they heard? Did they hear about the Kingdom, did they hear about love and hope, or did they hear about salvation? Different Christian denominations may present the "good news" in various ways.

Another aspect of this scripture is the meaning of "the end will come." Was this to mean within the lifetime of the hearers? Was He talking about the end of the world? Was He talking about the end of the Jewish system as it was known before Christ's time? Or, was He talking about the end of tribulation (as in Revelation) and the end of all things?

All these questions and more provoke thought, but none overshadow the core message here that we are indeed nearing the time when all the world will have heard the message of Christ. Keep in mind that we make this prediction from mathematical models and other supporting evidence. And, let it not be said that your author has joined the chorus of other false prophets who predict the end of the world. Only God knows!

Chapter 6

The Power of Prayer

What men usually ask for when they pray to God is, that two and two may not make four.

—Russian Proverb

Prayer can be described in many ways, among them as an intense desire to be in the presence of God, as communication with God, as petition with a fervent expectancy, and as a way to listen in order to grow spiritually. There is no one particular way to pray in form; many do it every day in many ways. Some let prayers flow from their mind; some read formal prayers—for example, Psalm 23— as a way to learn how to pray and to bring structure to their prayers. Some pray short prayers as they go about the day: "God, help me to bring your presence into my work, and my relationships with people." Some have prayers come to life between the lines as they read the Bible or a theology book, or listen to a sermon. Some find positive therapy by bringing any troubling issue to the forefront before God. Prayer can also be a kind of meditation where you use your mind to ask questions, and you listen for thoughts to come to mind that are creative, worthwhile, and unequivocal.

When prayer is offered to God for the benefit of another person, it is called *intercessory* prayer. If I pray for your safety or your recovery from illness, then I am offering an intercessory prayer. If I pray for my own safety or health, then I am offering a *personal* prayer.

C. S. Lewis, in his book *The Efficacy of Prayer*, wrote

What sort of evidence would prove the efficacy of prayer? The thing we pray for may happen, but how can you ever know it was not going to happen anyway? Even if the thing were indisputably miraculous it would not follow that the miracle had occurred because of your prayers. The answer is surely that a compulsive empirical proof such as we have in the sciences can never be attained.

Lewis died in 1963, but in 1988 and again in 1999, studies were done to evaluate the effect of intercessory prayer on the hospital stays of cardiac patients. The studies, mathematical and statistical in nature, and therefore scientific, contradict Lewis' claim that an empirical proof of prayer could never be attained. Prayer can be shown empirically to work.

After my heart attack, one area of Christian wisdom I became fascinated in was how prayer could be proven to work, especially where mathematics is involved. In a recorded sermon, Dr. David Jeremiah of Shadow Mountain Community Church in El Cajon, CA, referred to a medical study conducted by Randolph C. Byrd in 1988. Byrd is a Christian physician who at that time was in the Cardiology Division, Medical Service, San Francisco Medical Center, and the Department of Medicine at the University of California in San Francisco. I decided to delve into this study and another done in 1999 by William B. Harris. These studies showed a direct connection between improvement in patients (in this case, cardiac patients) and Christian intercessory prayer. To the extent that statistics is a branch of mathematics, these studies offered mathematical evidence for the power of prayer.

Dr. Byrd's Medical Study

Dr. Byrd's study involved 393 cardiac patients who were separated randomly into two groups. Over 10 months in 1988, one group

(PG, the prayer group) received intercessory prayer (IP), and the other received no such extra prayer (CG, the control group). Each patient in the study signed a letter of consent for the study, but after being assigned to a group, none of the patients knew whether they were or were not receiving the extra prayer. As a precaution against biasing the study, no consenting patient was ever contacted again, and none knew whether they were placed in the PG or CG.

Each patient in the PG was assigned to a group of three to five Christians for IP. The intercessors—the praying people—were chosen by the following criteria:

- They were "born again" Christians (followers of Christ) according to the Gospel of John 3:3.
- They had an active Christian life as manifested by daily devotional prayer.
- They were active in Christian fellowship, attending a local church.

Members of several types of protestant churches and of the Roman Catholic Church were among the intercessors. Daily prayers consisted of requests for a rapid recovery, requests for prevention of complications and death, and any other areas of prayer they believed to be beneficial to the patient. The intercessors knew the first name of the person they were praying for, but not the last name. They didn't meet the patients. It was assumed that some patients in both groups would be prayed for by people not in the study; this was not controlled for. The personal prayer on the part of each patient also was not controlled for. That is, there was no control or measure of how much, if at all, each patient prayed for themselves.

Neither the doctors, the patients, or the nurses knew whether a particular patient was in the PG (experimental) or CG (control). Statisticians refer to such a study as *double blind*.

Physicians can measure the effectiveness of a cardiac patient's stay in the hospital by using what is called a *severity* score, which involves the use of ventilatory assistance, antibiotics, and diuretics, as well as the frequency of congestive heart failure, instances of cardiac arrest, episodes of pneumonia, and instances that required intubation and ventilation (breathing assistance). The less each is used, the lower the patient's severity score. There was no statistical medical difference between the groups at the time of hospitalization, meaning that from the standpoint of medical measurements, they were the same in terms of the severity of their illness.

Prayer Group (PG)	Control Group (CG)
192 patients	201 patients

The results after 10 months? The PG had significantly lower severity scores on departure from the hospital. The CG required ventilatory assistance, antibiotics, and diuretics more frequently than the PG.

The conclusion was that intercessory prayer had an effect and it was presumed to be beneficial. Statistically speaking, the p-value for the statistical test was $p < 0.01$. PG did statistically better than CG, $p < 0.01$.

Statisticians understand the meaning of p-value and significant differences. It is their way of accepting or denying an hypothesis. Explaining it completely would take another book, but nonstatisticians can think of it in a simpler way. The statistical tests in the studies of this chapter use what is called *analysis of variance* (ANOVA). A researcher tests hypotheses, called *null hypotheses*. Let's consider the test for the prayer factor. The null and alternative hypotheses (in words) are as follows:

Null Hypothesis: There is no effect due to the prayer factor.

Alternate Hypothesis: There is an effect due to the prayer factor.

If we can reject the null hypothesis (that is, accept the alternative), we say that the test is *statistically significant*. Another way of saying this is that the value of the test statistic is significantly different than what we could reasonably expect if the null hypothesis were true. By *significantly different*, we mean that the probability is low that the difference between the value of the test statistic we observe and the value we would expect to observe (if the null hypothesis were true) is purely due to chance. If the difference is statistically significant at the 5 percent level, the probability that the difference is due to chance is equal to or less that 0.05; if the difference is statistically significant at the 1 percent level, the probability is equal to or less that 0.01; and so on.

The thought process concerning the concept of statistically significant difference is nonintuitive. It may be why so many students hate statistics! To sum up as simply as possible, there is a statistically significant difference if the probability that the observed difference is due purely to chance is extremely low. Interpreted in terms of the Byrd study, there was a statistically significant difference between the PG and CG; that is, the probability was less than or equal to 1 percent that the difference was due to chance. To a statistician, Byrd's study proved intercessory prayer was effective.

Dr. Harris' Medical Study

Medical researchers—indeed, all statistical researchers—value replication of an effective study. In 1999, Dr. William S. Harris and his team set out to replicate Byrd's study on the power of prayer. Harris is a Christian nutritionist and was assisted by several physicians. The study, administered to cardiac patients at the Mid America Heart Institute in Kansas City, MO, over a 12-month period, was carried out much the same way as Byrd's, but with more patients (1,005). This time, however, neither the prayer nor control group knew the study was being done. No one knew prayer was being studied except the experimenters and those doing the praying, and there was no difference in the severity scores of the PG and CG at the time of admission to the hospital. The study, again double blind, showed that the PG later had a significantly lower severity score than the CG upon departure from the hospital.

Prayer Group (PG)	Control Group (CG)
466 patients	539 patients

The study found the PG did statistically better than CG, $p < 0.01$.

Comparison and Criticism of the Studies

In an interview with me in 2005, Dr. Byrd commented that he wished he could have included a larger number of patients in his study. Harris achieved that goal by having 1,005 people in his study versus 393 in Byrd's study. Statisticians have a high regard for studies with big numbers (big n values), so it was an additional validation that the study with higher numbers yielded similar results. The other primary variation between the two, aside from sheer numbers, was that in Byrd's study, all the patients consented to the study, although each patient never knew whether they were in PG or CG. In Harris's study, the patients had no idea they were part of the study.

Both Byrd and Harris endured personal and professional assaults for doing their studies. Colleagues and Christians alike raised questions such as, "How can you test God in this way?" Once, Harris was verbally attacked and thrown out of his minister's office. What was the minister really afraid of? Was it his own dread of discovering scientific evidence of something that he had taught all of his ministerial life? Why should he fear new insight from science and from God?

A physician criticized the Harris study, asking, "Why should God allow the patients who received the remote intercessory prayer to do better than the control group? Does God love those for whom strangers pray more than those who were randomly assigned not to receive their prayers? I was taught that God is not capricious and that faith is not a matter of scientific proof." In fact, it was neither a goal nor a result of Harris' study to answer these questions.

Harris and his group responded as follows:

A critically important attribute of any scientist is open-mindedness, the willingness to objectively consider new or alternative concepts and hypotheses. There is a growing demand among patients that we acknowledge their need to be treated as whole persons [using holistic medicine] who have not only physical but emotional and spiritual needs as well. Practicing as we do in a large metropolitan hospital among a wide variety of religious traditions, we are acutely sensitive to the need for a nonsectarian approach to addressing spiritual issues. This diversity is mirrored in the spectrum of religious practices among our doctors, which ranged from a variety of Protestant and Roman Catholic traditions to Hinduism. Since spiritual factors may play some role in healing, additional studies are needed to clarify the place of intercessory prayer in maintaining and restoring health.

Why should people, especially Christians, be so afraid of wisdom and knowledge drawn from these careful, well-done studies? Dr. Harris, who also consented to a telephone conversation, told me that he had resolved to publish the results of his replication no matter the results: negative, neutral, or positive.

Harris commented on what his study did not show, saying

We have not proven that God answers prayer or that God even exists. It was intercessory prayer, not the existence of God, that was tested here. All we have observed is that when individuals outside of the hospital speak (or think) the first names of hospitalized patients with an attitude of prayer, the latter appeared to have a "better" cardiac care experience.

When I spoke to Dr. Harris about this statement, he raised the question "How do you define God?" I had no simple answer, taking as my faith axiom that God is undefined but very real to me. He raised the issue pertaining to both studies that the intercessors knew only the first names of those they were praying for. How did God know the difference between a man named Todd who was a cardiac patient in PG versus a man named Todd in CG, let alone others by that name in the hospital at that time? Somehow, I think God knew.

Studies on Prayer Reporting Nonsignificant Differences

Although we'd all love to believe that prayer can always make a difference, other scientists studying the efficacy of Christian prayer have sometimes found other results. In embracing both sides of this paradox, we have found that some studies did not result in significant differences. For example, in 1997, Dr. S. O'Laire found no significant differences in the effect of intercessory prayer on self-esteem, anxiety, and depression in 406 subjects (who received either no prayer, directed prayer, or nondirected prayer), using 90 intercessors. In 1997, Dr. S. R. Walker and his colleagues at the University of New Mexico found no specific benefits on 40 recovering alcoholics.

Another prayer study that did not result in significant differences was done by Mitchell Krucoff, Suzanne Crater, et al at nine centers throughout the U.S. and published in *The Lancet*, a well-respected British medical journal, in 2005. Patients receiving heart surgery for unstable coronary symptoms were randomly assigned bedside intervention using music, imagery, and touch therapy or no intervention, and sites were informed of the assignment. Patients were simultaneously randomly assigned off-site prayer or no off-site prayer, but sites and patients were not informed (double blind). Telephone randomization-center services were provided by Interactive Clinical Technologies, Inc. (Durham, NC) and the Duke Clinical Research Institute. In short, no significant differences were found. But an important point in this case is that each prayer group worked in its own geographic location and within its specific religious faith. Each group was independent. From the patient side, however, it was the reverse. Each patient who was assigned the prayer treatment received healing prayers from all prayer groups. That is, each patient who was assigned the prayer treatment received healing prayers from mixtures of prayer groups, even though each group consisted entirely of one of the various religions: Christian, Muslim, Jewish, and Buddhist.

Herbert Benson, Jeffrey Dusek, et al of Harvard Medical School did a study published in 2006, in which patients about to undergo cardiac bypass surgery received intercessory prayer. The study consisted of a PG and a CG, but also a third group PG'. The PG and CG groups were assigned randomly with no person in either group knowing for sure whether they received prayer. But the PG' group were told that they would be receiving prayer. After 30 days of follow up, no significant differences were found between the three groups on measures of major events, mortality, or complications. When the PG and PG' groups were compared, it was found that the PG' were 14 percent more likely to experience complications. Although this difference was not significant, it is fascinating to note that those who were uncertain whether they were receiving prayer did better than those who knew. That is, receiving prayer had a negative outcome, though again it was not a significant difference. The best explanation your author can come to for this happening is the influence of "pressure to perform" on the patients. Send me your thoughts.

Medical journals tend to place higher value on studies that result in significant differences. It is possible that there exist many other studies with nonsignificant differences that lie unpublished on the desks of medical researchers.

Other Studies on Prayer Reporting Significant Differences

Approximately five other studies besides those of Byrd and Harris have been performed that reported significant differences. For example, Fred Sicher et al of the Geraldine Brush Cancer Research Institute (California Pacific Medical Center, San Francisco) in 1998 found statistically significant benefits for the intervention group (fewer new illnesses, physician visits, hospitalizations, and days of hospitalization; lower illness-severity scores; and improved mood scores. A study by Leslie Furlow and Josie O'Quinn in 2002 reported significant differences in favor of intercessory prayer in 21 primarily male, elderly, Christian cardiac patients, although the sample was small.

Another example comes from a study by Leonard Leibovici in 2001 at the Rabin Medical Center (Petah-Tiqva, Israel). You think God does not have wonders to perform and can't work in *reverse*? Well, Leibovici's significant results will astound you. Leibovici's study showed that intercessory prayer worked for patients well after they had left the hospital. The study analyzed a group of patients from the standpoint of length of stay in the hospital and duration of fever from blood-stream infections. The group was split into a PG and a CG, and the PG group received intercessory prayer, although in hindsight. The study reported significant differences for the PG group.

A Controversial Study

For the skeptics in the crowd, it should be noted that there has been at least one study on prayer and medicine that is embellished in controversy. The study cited has to do with fertility. In September 2001, *The Journal of Reproductive Medicine* (JRM), a highly respected medical journal, published a study conducted at Columbia University by Dr. Kwang Cha, Daniel Wirth, and Rogerio Lobo. They wanted to examine whether prayer had an effect on the success of in vitro fertilization (IVF) procedures. Their contention was that those in the prayer group had a higher pregnancy success rate: 50 percent versus 26 percent.

However, the study came under attack in articles by Paul Harris in *The Observer* and by Jacob Gershman in *The New York Sun*. The articles discussed the efforts of Dr. Bruce Flamm, a clinical professor at the University of California and metaphysical atheist, who tracked and refuted the study. To increase the success rate of IVF by 100 percent would be "a huge breakthrough, a revolution," according to Flamm. But Wirth pleaded guilty in 2004 to multimillion dollar fraud charges and has used a series of false identities over the years. Wirth also has no medical degree. Cha has left Columbia but would not answer phone calls or emails sent by Paul Harris. This is in contrast to the Byrd and

Harris prayer studies, where both researchers were easily contacted by me. In order to do further investigation, the JRM removed the discredited study from its site for a period of time but after further analysis returned it to the site, but deleted Lobo's name.

Dr. Hodge's Research Review of Empirical Literature on Intercessory Prayer

Dr. David R. Hodge, a professor of social work at Arizona State University, published a document in 2007 in the journal *Research on Social Work Practice* that reviews a vast amount of empirical literature regarding prayer. At the least, Hodge's paper supplies us with an encyclopedia of references for those seeking information about studies regarding prayer. At the most, it reports the results of an advanced method of statistical research, called *meta-analysis,* which allows the results of several studies to be combined to draw a positive conclusion regarding the power of prayer. Our focus in this book does not allow in-depth mathematical development of meta-analysis, which is well regarded by statisticians. Meta-analysis must be used with some caution because of the way it combines results. Nevertheless, it can reveal worthwhile results. For example, six different studies indicated nonsignificant differences regarding the use of aspirin to prevent heart attacks. Yet when the studies were combined using meta-analysis, aspirin was found to yield significant differences and is now a widely accepted preventative.

Although Hodge lists more than 75 studies in his paper, he limited his meta-analysis to 17 studies, with 7 reporting significant differences and 10 reporting nonsignificant differences. The controversial Cha study was included in the meta-analysis. The result of the meta-analysis showed that prayer produced statistically significant results. The conclusion for Hodge is that the balance of the scale tips on the side of prayer: Prayer works.

What Does All This Mean?

What are some of the personal conclusions I draw from these studies, especially those of Byrd and Harris and the meta-analysis of Hodge? Quite simply, prayer works. If 2,000 studies resulted in positive results regarding prayer, there would still be controversy. In Rev. Jeremiah's sermon, he closes his reference to Byrd's study with a quote from Larry Dossey, in which he asserts, "Physicians who don't pray for their patients are guilty of malpractice." Regardless of whether you feel that may be going too far, it would be gratifying to see more medical professionals acknowledge the power of prayer. Personally, I'd like to see the results in the headlines of every newspaper, but they should at least make the front page of a medical journal.

There are a variety of ways I believe we can mathematically, statistically, or scientifically uphold the belief that prayer works. One way is to cite the mathematical and statistical approach Harris and Byrd used in their studies. Suppose you are a having a debate with someone regarding the power of prayer, and you are trying to present an apologetic argument. You tell them stories about how convinced you are that prayers are answered, only to have them dismiss your observations with the comment, "Your answers to prayers were just chance happenings. What about the times your prayers were not answered?" Two questions present themselves here. God always answers, but in His own way—maybe not exactly when you ask, or in the way you want. Sometimes we have to wait, sometimes the answer is "No," and sometimes the answer far exceeds what we ask for.

The response to the idea of "chance happenings" is to ask the person with whom you are engaging in an apologetic argument about their belief in science. If they believe, then you might ask whether they believe the statistical studies on the negative effects of cigarette smoking. Most people have faith in these scientific results. If you share the results of the prayer studies with them, then they have to accept them based on their acceptance of other scientific results such as the smoking studies. Both studies were performed in scientific conditions using similar methods and drawing statistically significant differences from the data.

A Possible Follow-up Study on Prayer

A follow-up prayer study could involve examining the righteousness of the person praying and the fervor of the prayer. Are the results more pronounced when someone prays with more sincerity? A cardiologist and friend of mine, Dr. Bruce F. Schilt, who reviewed the medical results of the Harris and Byrd studies, suggested a follow-up, referring to the following Biblical quote:

> **James 5:16:** *Confess your sins to each other and pray for each other so that you may be healed. The earnest prayer of a righteous person has great power and wonderful results.*
>
> Here the word *righteous* means "the state of being without guilt or sin." A Christian becomes righteous by praying for forgiveness from sin.

Dr. Schilt commented that if these studies were "pills, or medications" that a pharmaceutical company were testing, then the next step would be to test them on a wide basis to an expanded population, again using a carefully controlled scientific study; only this time, the population size would be much larger. In the case of prayer, humankind has been testing and proving its power for centuries, but the results have been conveyed in anecdotal form rather than in controlled, large sample scientific studies. I welcome these kinds of studies and look forward to their results.

Conclusion

The claim by C. S. Lewis that "a compulsive empirical proof such as we have in the sciences can never be attained" seems to be refuted by the two medical studies of Byrd and Harris and others. These studies, based on mathematics and statistics, at the very least conclude that prayer does have a positive effect on the hospital stays of cardiac care patients. The mathematics provides an apologetic defense of the power of prayer. But prayer is not always directed toward medical cures. Many prayers of petition seek material possessions, wisdom, security, and direction; others seek communication and the desire to possess God and be in relationship with Him. In the sense of these types of prayers, conclusive empirical proof remains to be discovered.

Chapter 7

Higher Dimensions: Metaphor for Spiritual Insight

We cannot see light, though by light we can see things. Statements about God are extrapolations from the knowledge of other things which the divine illumination enables us to know.

—C. S. Lewis, *The Four Loves*

A *metaphor* is a figure of speech in which a word or phrase such as "A Mighty Fortress Is Our God" is used to describe another idea. Or, it may be used to make an implicit, or parallel comparison, as in "hiking in Canyonlands National Park in Utah is like walking in heaven." Edgar Allen Poe's short story "The Tell-Tale Heart" is an example of a metaphor in literature. In that story, the narrator murders a man and takes pride in hiding the body by cutting it in pieces and concealing them under the floorboards of his home. But the murderer is haunted by his deed when he repeatedly hears the beat of the victim's heart. The guilt that arises from the continual, taunting heartbeat eventually leads the murderer to confess his crime to the police. The story can be a metaphor of the guilt we all feel when our conscience condemns a wrong we have done and we seek redemption.

Christ's parables are excellent examples of metaphors. They are often referred to as earthly stories with heavenly meanings. Consider, for example, the story of the woman at the well in John 4. Christ was passing through Samaria and stopped to sit at a well. A Samaritan woman stopped, and He asked her to get Him a drink. No Jew would have anything to do with a Samaritan, let alone a female Samaritan. In shock, she asked, "Why are you asking me for a drink?" His response was, "If you only knew the gift God has for you and who I am, you would ask me, and I would give you living water ... the water I give them [meaning anyone] takes away thirst altogether. It becomes a perpetual spring within them, giving them eternal life." Christ's metaphor compared the water of the world with His "living water," which is actually "heavenly wisdom."

In the first part of this book, we focused on analytical mathematics. In the next few chapters, we use imagination and mathematics to create a metaphor for spiritual insight through mathematical concepts of higher dimensions. Most people know of four dimensions: length, width, height, and time. In everyday life, we may extend these by considering other attributes such as color, mood, taste, smell, and so on. Each adds a higher dimension to our reality. But scientists consider other, more sophisticated dimensions, such as mass, density, force, and speed. Mathematicians consider dimensions as coordinates by listing pairs of numbers (x_1, x_2), triples of numbers (x_1, x_2, x_3), n-tuples of numbers $(x_1, x_2, x_3,..., x_n)$, and so on. Details of the scientific and mathematical concepts will emerge as we proceed.

As a metaphor, we will be using the nth dimension to imagine the $(n+1)$st dimension of heaven and the Existence of God. Put on your Technicolor imagination hat, bring your mathematics, and savor the mystery!

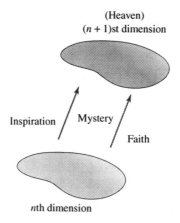

Figure 7.1 Moving from the nth dimension to the $(n+1)$st dimension is a progression involving imagination, mystery, and faith.

Higher Dimensions as a Metaphor for the Soul

A *dimension* can be thought of simplistically as a property that describes an object. We will enhance this concept in the rest of this chapter. A simple beginning model of the progression from lower to higher dimensions is a series of planes stacked like stairsteps.

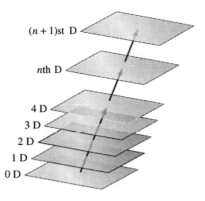

Figure 7.2 Moving from one dimension to the next can be imagined as a series of planes stacked like stairsteps.

In his book *The Weight of Glory*, author C. S. Lewis describes a situation where a pregnant woman is placed in a prison cell with a "little patch of sky seen through the grating, which is too high up to show anything but the sky." She is an artist and has a drawing pad and a box of pencils. After a son is born, she tries to teach him about the outside world by means of pencil drawings. She draws lots of pictures of rivers, mountains, cities, waves on a beach, and so on. Eventually, they are released, and the boy says, "What? No pencil marks there?" He cannot imagine a world without pencil marks. No matter how hard she tried, her pencil drawings were inadequate to explain the outside "higher" dimension.

We normally think of a three-dimensional object as having properties of length, width, and height. But other dimensions can be considered, such as the time at which an object is thought of, as well as its mass, volume, and color; or, in the case of a living human, the person's gender, mood, ethnicity, religion, or political stance. Consider my oldest son, Lowell D. Bittinger, as an example. Lowell is 6 feet, 5 inches tall, with dark brown hair; he's a loving son, father, and husband, an actuary by profession who enjoys basketball, baseball, football, and playing golf. Of all the members of our family, Lowell is the most patient and understanding. The more I say to describe Lowell, the more dimensions I add to his character.

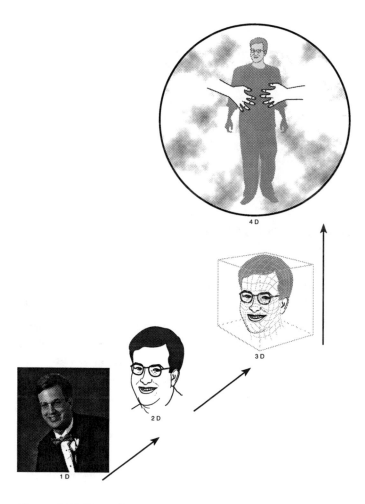

Figure 7.3 Four dimensions of Lowell D. Bittinger

You can see four dimensions of the same person, my son Lowell, although the last three require your imagination. The first dimension (1D) is a black-and-white photo. The second dimension (2D) requires your imagination to think of a full-color photo of Lowell. What if Lowell walked into the room? Use your imagination of this event for the third dimension (3D).

Think over this sequence again, but this time suppose you live in a world that consists entirely of black line art (0D). I'm sitting next to you, trying to describe the notion of a black-and-white photo (1D). You can see certain aspects of Lowell in the line drawing. But try as I might, with all the creativity I can muster, I always come up short of constructing the total concept of a black-and-white photo. Can you imagine the thrill of finally entering the world of black-and-white photos?

Now, suppose you live in a world that consists entirely of black-and-white (1D) photos. I'm sitting next to you, trying to describe the notion of full color (2D). Try as I might, with all the creativity I can muster, I always come up short of building the total concept of a full-color photo (2D).

Suppose you exist in a world consisting of only full-color photos of people. I try with all the creativity in my mind to describe what a live person is like, but you just can't get it. Lo and behold, Lowell walks into the room, and you meet him. Not only do you get to interact with the live Lowell, but you become aware of other three-dimensional objects as you talk to another person and touch and interact with objects.

Do you see where we are heading? Think of the "soul" of a person as a higher dimension (4D). Someone in a higher dimension is trying to tell us about the idea of you and me each having a soul. How might He go about it? We can read descriptions in the Bible. We can get a sense of the soul in our relationships with other people and in the good acts they perform. We can a sense of the soul by listening to ministers' sermons and by reading theological books of Godly writers. You might read that the soul is one of life's mysteries. It is the power of life within you; it consists of your thoughts, actions, and emotions. The soul is the spiritual entity inside you, separable from the body at death, that transcends this life to the next. Such descriptions always fall short.

Imagine Lowell standing next to you, trying with all his power to convey to you his next dimension, his *soul*. Human beings struggle to know their own souls, even though each person has the advantage of awareness of their own inner thoughts. It's inadequate to use a lower dimension to describe or understand a higher dimension. The real Lowell gives us many ideas about his soul, but only God—not even Lowell in this lifetime—knows his own soul. Try as I may, using all the dimensions available, I can never totally describe the soul of Lowell or another person. In fact, I am in the same boat as you. I have no adequate means of describing your soul or mine. But, we do have ideas or rumors to go on. Can you imagine the joy of entering a world where you can experience your soul or that of another person? Could that be similar to being in heaven? Can you embrace the mystery that will be revealed at that time?

Joseph Campbell wrote that "People say that we're searching for the meaning of life. I don't think that's it at all. I think that what we're seeking is a spirit of being alive, so that our life experiences on the purely physical plane will have resonances within our innermost being and reality, so that we actually feel the rapture of being alive." Lowell's "soul" is an example of the "rapture of being alive." In summary, the nth dimension, no matter how much we learn about its characteristics, is always inadequate to describe the $(n+1)$st dimension or higher. We live in this life with the mystery of the soul, even though we frequently encounter hints or rumors about it.

Higher Dimensions as a Metaphor for Heaven

Are you getting the idea of using your imagination together with the notion of higher dimensions to conjure some of life's mysteries? Let's consider higher dimensions as a metaphor for another key mystery of life: heaven. What is heaven like? Scientists, philosophers, authors, movie makers, musicians, and artists have tried with all their creativity to describe heaven. Some people think heaven will be a return to the

Garden of Eden before Adam and Eve ate the apple from the tree of knowledge of good and evil. M. Scott Peck thinks heaven will be a situation of community where people love each other in perfect harmony.

My son Lowell thinks of lights, gold, peacefulness, and lots of pretty music. My youngest son, Chris, thinks of perfect love and becoming perfect—that is, being whole and faultless. He also thinks of heaven as a place where you can live out how God made you in a perfect way. My wife, Elaine, thinks of the utter awesomeness, which is incomprehensible to us in our finite existence.

I have several notions of what heaven could be like. The first idea, although tongue-in-cheek, is heaven as a place where I can eat anything I want: pancakes with real maple syrup, double cheeseburgers, meat loaf, sloppy joes, baked beans, corn, jello, and red-raspberry pie covered with vanilla ice cream. (I am literally able to eat an entire pie at one sitting.) Such are my thoughts in this life as I endure a low-fat, low-calorie diet for the sake of my health. (Sometimes, I feel I'm in an almost constant state of semi-hunger.) My second notion of heaven, more profound, involves seeing family and loved ones who have passed away—especially my mother, whom I have never met, and then my father and the grandmother who raised me.

The third notion is sitting down with God and asking all the questions that remain unanswered in this lifetime. For example:

- What is the real story behind the assassination of John F. Kennedy?
- Can you show me a videotape of the Big Bang?
- What really happened on Flight 93 on 9/11/01?
- Is there life on other planets, and if so, can I see it?
- Why did you bother to make us?
- Did somebody make you?
- What do you think about mathematics?
- Can you show me a list of all the math theorems, and allow me to understand them?
- Why do children have to be born with birth defects?
- How did my mother die 10 days after my birth? Did she commit suicide, or did she fall from that hospital window by accident?

- Have I lived the kind of life you wanted, to resolve the paradox of her death?
- Will you answer 1,000,000 more questions?

Other wishful ideas about heaven follow:

- I will still be able to hike near my beloved Moab, Utah, but God will have graced the experience in his $(n+1)$st dimension beyond anything I can expect in this lower nth dimension.
- I will be able to eat anything I want to my heart's desire without any health repercussions.
- I won't have to run on the treadmill anymore in order to extend my life expectancy.
- I will be reunited with all my friends and loved ones in a loving fellowship of community.
- All my mathematical questions will be answered.
- If I get to teach math again, all my students will be eager and well-prepared and will desire to integrate math into their professions.
- Elaine, my wife, will still be my loving companion. We will travel, hike, explore, and love all our acquaintances as Christ has taught us.
- All my paradoxes will be resolved!

I now ask my reader to stop and make a list of what you think heaven might be. If you are reading this book in a group, share your lists.

Once you've thought about what you think heaven might be for you, let's use the metaphor of higher dimensions to imagine it. Using a lower dimension to describe a higher one is grossly inadequate. Nevertheless, I will reveal my personal attempt to describe heaven using imagination and mathematical experience.

Let's take, for example, my first idea, which involves hiking in the Needles District of Canyonlands National Park near Moab, Utah. One of my favorite spots is close to a formation I lovingly call "Marv's Monolith." To describe heaven to you, I might start with a black-and-white photograph and then move to a higher dimension with a full-color photograph. To move to a yet higher dimension, I would hike with you to the spot and share with you the splendor of the area. I often listen to my favorite music on an iPod while I hike. There are also special times when I stop to read and think. I love the red, orange, brown, and yellow in the formation and in the panorama around it. I am in total wonder of how anything that unique can exist. I appreciate the quiet of that spot and the typically dark, clear blue sky in the background. I am from a state whose geography is mostly green and flat; it has a beauty of its own, but none of it jumps up and grabs me like this monolith. Only God could have made it. At these special moments, I have the feeling that I am as close to heaven or God as I can get in this lifetime.

Figure 7.4 Marv's Monolith

We started our thinking with a black-and-white photograph, which we can consider the first dimension. Then we moved, as before, to a full-color photograph as the second dimension. Now, let your mind conjure a description of a full-color photograph using black and white. How do you imagine the next dimension? Some ideas would be to move to an improved color photo taken with a high-resolution camera and enhanced with Adobe Photoshop®. Another addition is to play music. But no matter what you can imagine, using a lower dimension to describe a higher dimension is woefully inadequate. Let's try to imagine the next higher dimension to be heaven.

To get an idea of how daunting this task might be, consider the creative job of being a music composer writing a symphony. The composer, in almost a God-like position, has a concept of how the entire piece should sound, but he has to break down the composition for the individual instruments. He writes the symphony for the trumpets, then the violins, then the drums, and so on. Now, try to think the other way, considering each instrument independent of the others. The composer has heaven in mind—his composition—but each instrument is inadequate to convey the entire piece.

Think of heaven as the $(n+1)$st dimension and the dimension we live in as the nth dimension. We cannot totally imagine the $(n+1)$st dimension. To visualize heaven, we must stylize all the perceptions of the nth dimension we know. I can reframe all the thoughts, all the words, and all the dreams I have in the nth dimension. Even if I talk in ethereal language, even if I talk in an other-worldly language of what heaven is like, it's still only the nth dimensions that I can fully understand. I cannot talk about heaven if it is the $(n+1)$st dimension. I can only use the notion developed here that moving from one dimension to the next is amazing, overwhelming; a jump beyond what my mind can imagine; a leap of "faith." I can tell you unequivocally that I can't wait for the mystery to be revealed!

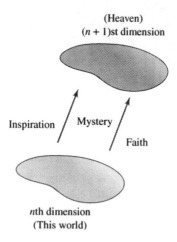

(Heaven)
$(n + 1)$st dimension

Inspiration / Mystery / Faith

nth dimension
(This world)

Figure 7.5 Moving from the nth dimension to the $(n + 1)$st dimension is a progression involving imagination, mystery, and faith.

Higher Dimensions and Mystery

We now use higher dimensions to build up to an nth dimension. But, here it is not so clear what we are building up to—a certain amount of mystery will linger. We can think of the concept of higher dimensions using a variety of conceptual tools or metaphors. Music is one of my favorite examples. First, pick any piece of music you love. It can be classical or modern. One of my favorite pieces is the theme from the movie *Ice Castles,* "Through the Eyes of Love." Consider the following sequence of dimensions, moving from 1D to 2D to 3D to 4D and higher.

The nth dimension, no matter how much we learn about its characteristics, is always inadequate to describe the $(n+1)$st dimension or higher. Suppose you can see a piece of sheet music (1D), but you are either deaf or have no concept of sound. You might feel vibrations or watch someone playing a piano, but try as I might, the first dimension is inadequate to describe the second. Then, you do hear the piano music. What an amazing revelation! You become comfortable with many kinds of piano performances. Now, I come up to you and start telling you that there is a better dimension (3D): an orchestral rendition. But the only thing you can hear is a piano playing a piece of music. You can hear the beat and the flow of the music and maybe know whether you like it, but the lower dimension (2D) never totally conveys the beauty of an orchestral rendition (3D). Then, you hear an orchestral rendition of the song "Through the Eyes of Love," from the movie *Ice Castles.* A thrilling new dimension is revealed to you.

There is a touching scene in the movie when the figure skater Lexie, played by Lynn-Holly Johnson, has become blind; but thanks to the encouragement of her boyfriend Nick, played by Robby Benson, she returns to the ice and wins a contest. As this happens, a full orchestra plays the theme song. You see this clip in the DVD and think of a fourth dimension (4D). In each case, the nth dimension gives clues, hints, or rumors about the $(n+1)$st dimension but is inadequate to describe it completely.

1D: Sheet music. You live in a world consisting of sheet music. No sound! You examine a piece of sheet music for "Through the Eyes of Love," the theme from the movie *Ice Castles*.	I'm sitting next to you trying to describe the notion of hearing the music come to life on a piano (2D). Try as I may, with all the creativity I can muster, I always come up short of building the total concept of live piano music. Can you imagine the thrill of finally entering the world of live piano music?
2D: Piano music. You live in a world consisting of piano music, with no other instruments or orchestral sounds. You hear a piano recording of the "Through the Eyes of Love," the theme from the movie *Ice Castles* performed by Emile Pandolfi, on the CD *Emile Pandolfi by Request*, MagicMusic Productions, Greenville, SC. (Through the Eyes of Love) By Marvin Hamlisch, Gold Horizon Music Group (ASCAP).	I'm sitting next to you trying to describe the notion of hearing the piano music come to greater life through an orchestra (3D). Try as I may, with all the creativity I can muster, I always come up short of building the total concept of live orchestral music. Can you imagine the thrill of finally entering the world of live orchestral music?
3D: Orchestral music. You live in a world consisting of orchestral music. No videos or DVDs exist to enliven the music with a movie. You hear an orchestral recording of "Through the Eyes of Love," the theme from the movie from *Ice Castles*, performed by Erich Kunzel directing the Cincinnati Pops, on the CD *Movie Love Themes*, Copyright© 1991, TELARC®, Cleveland, OH.	I'm sitting next to you trying to describe the notion of hearing orchestral music and seeing a DVD movie (4D). Try as I may, with all the creativity I can muster, I always come up short of building the total concept of live orchestral music as a part of a movie. Can you imagine the thrill of finally seeing the movie and hearing the music?

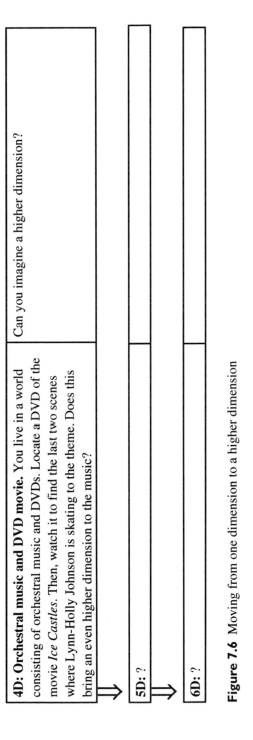

4D: Orchestral music and DVD movie. You live in a world consisting of orchestral music and DVDs. Locate a DVD of the movie *Ice Castles*. Then, watch it to find the last two scenes where Lynn-Holly Johnson is skating to the theme. Does this bring an even higher dimension to the music?

Can you imagine a higher dimension?

5D: ?

6D: ?

Figure 7.6 Moving from one dimension to a higher dimension

But what might be the fifth dimension? Based on the pattern of the first four dimensions, stop and let your mind wonder about the possibilities and the mystery. If I were speaking to you in a group, I would pause and allow time for your mind to wander, or for people from the audience to make suggestions. One notion that may come to mind is some yet-to-be-invented way of enhancing the music. There are many possibilities, but because we don't have the convenience of sharing ideas, I'm going to give you mine.

Have you seen movies such as *It's a Wonderful Life, Braveheart, High Noon, The Magnificent Seven, Rocky, Cinderella Man, Legends of the Fall*, or *Ice Castles*? They have great audience appeal because all of them are remarkable stories of love, heart, or redemption after a tragedy. But I watch such movies and yearn to move them to a next dimension. I'd like to see the characters portray what they do with a God-like or Christ-like consciousness. Then those movies would enhance God's plan for the hearts of the movie characters, and the audience would come closer to sensing God's glory. However, most Hollywood moviemakers would never make such a bold move away from a secular consciousness.

Let's illustrate this with the movie *Ice Castles*. In the film, the characters have no overt religious affiliation; no scenes take place in a church, and there is no sense that Lexie has been blessed with her skating skills by God or that she has in her heart the desire to use them to please God. Before her blindness, she moves into a world of celebrity and self-centeredness that frustrates her. That could have come from a feeling of discomfort she had due to getting away from a Godly consciousness, but that opportunity was lost. In her frustration, she leaves a party and goes to do what she does best—skate in a nearby rink—only to slip on some power cords, hit her head, and lose most of her eyesight. She does redeem the tragedy, but through no faith in God—at least, not that the audience is aware of. Her faith and that of Nick, who helps her, could have been exemplified in this movie. Christ could have been shown to come into them and work through them to reveal His love and hope—His glory.

Using Metaphor to Describe the Existence of God

From a philosophical standpoint, I accept God as a faith axiom. I arrive at that faith axiom by considering the higher dimension of many types of situations, based on the following Biblical quote of the Apostle Paul:

> **Rom 1:20:** *From the time the world was created, people have seen the earth and the sky and all that God made. They can clearly see his invisible qualities—his eternal power and divine nature. So they have no excuse whatsoever for not knowing God.*

The following are some situations in which my imagination and faith have allowed me to reach from the nth dimension of this world to the $(n+1)$st dimension of the existence of God:

- The birth and growth of a child. As I've watched my grandchildren grow, I'm constantly amazed and only now imagine the higher dimension that created them to be God.
- The Utah scenery. Only now, I imagine the higher dimension to be the maker of that scenery, God. (For others, nature settings of a seashore, a desert, or a jungle may allow imagination of God.)
- Great works of music, such as operas and symphonies, or even bodies of work by popular singers such as Luther Vandross or Allison Krauss, or composers such as Marvin Hamlisch.
- Great works of literature that have touched and inspired millions of people over hundreds of generations by authors such as C. S. Lewis, Fyodor Dostoyevsky, Philip Yancey, and Brian McLaren, and especially translations of *The Holy Bible*.

- Great works of art created by artists who can inspire beauty, emotion, and feeling in paintings, drawings, sculpture, and many other media.

Many of us see evidence of God's existence around us every day, in

- A beautiful garden bursting with flowers in bloom, and the seeds from which they grow
- The incredible workings of the human body, especially the human eye, the heart, and the brain
- A bowl of fresh red raspberries, or a raspberry pie topped with a pint of vanilla ice cream
- Meat loaf from a good restaurant, a double cheeseburger, or a Thai chicken pizza
- A stack of pancakes drenched with real amber maple syrup from Vermont, with an aftertaste of nature
- Words of the Bible
- Rainbows
- Meaningful sermons
- The power of prayer
- Having one of my granddaughters, as a baby, fall asleep on my chest, and listening to her breathe
- The romantic love of my wife

The union of all the imagined higher dimensions of these situations provides proof to me of the existence of God. "See the nth dimension, imagine the $(n+1)$st dimension," or "See the gifts, seek the giver!"

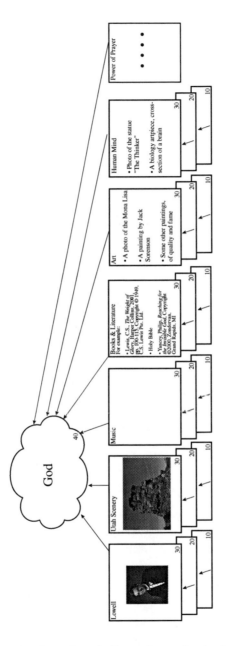

Figure 7.7 Moving from the *n*th dimension to the (*n*+1)st
dimension: a progression involving imagination, mystery, and faith

God: Intelligent Designer of the Universe

I have used higher dimensions as a metaphor to argue for the existence of God. The *theory of intelligent design (ID)* holds that certain features of the universe and of living things are best explained by an intelligent cause that we call God. Let's examine some arguments for the existence of God as intelligent designer. In so doing, we enhance the metaphor argument.

Please note that I am considering the notion of intelligent design here just as I have described it. ID has gotten a great deal of press as standing at the opposite of a debate regarding evolution. I will not be considering evolution in this book. For us, here, in relation to the metaphor, I'm talking about the concept of God strictly as the Intelligent Designer (as opposed to the entire, sometimes controversial and misunderstood, idea of ID versus evolution). Some people believe that an intelligent God made our genetic code adapt to conditions and change based on survival. I will not pursue such matters. I use the dimension metaphor to argue that there was an Intelligent Designer of the universe, which I call God.

A more commonly read philosophical case comes from William Paley, eighteenth-century British theologian-naturalist; it was his Watchmaker Argument:

In crossing a heath, suppose I pitched my foot against a stone, and were asked how the stone came to be there; I might possibly answer, that, for anything I knew to the contrary, it lain there for ever: nor would it perhaps be very easy to show the absurdity of the answer. But suppose I had found a watch upon the ground, and it should be inquired how the watch happened to be in that place; I should hardly think of the answer which I had before given, that for anything I knew, the watch might have always been there.... The watch must have had a maker: that there must have existed at some time, and at some place or other, an artificer or artificers, who formed it for the purpose which we find it actually to answer; who comprehended its construction, and designed its use.... Every indication of contrivance, every manifestation of design, which existed in the watch, exists in the works of nature; with the difference on the side of nature, of being greater or more, and that in a degree which exceeds all computation.

According to Hugh Ross in his book *The Creator and the Cosmos*, "...all the naturalists of Paley's day admitted and all the biologists of today emphatically concur [that] the complexity and capability of living organisms far transcends anything we see in a watch. If a watch's complexity and capability demand an intelligent and creative maker, surely, Paley reasoned, the living organisms on our planet demand a Maker of far greater intelligence and creative ability." God was the Maker, the Intelligent Designer.

Another metaphorical argument we can use, which is similar to the Paley argument, is one of perspective. Although it's not about aliens creating anything on our planet, as some propose for the pyramids, Stonehenge, and the giant heads on Easter island, it can be about what an alien would think if he, she, or it landed on the Moon tomorrow and on Earth the next day. Suppose you were from an alien culture and had traveled toward Earth, stopping first on the Moon. Keep in mind, you have to be of high intelligence to travel through space to the Moon. You discover the remnants of man's travel to the Moon: the lunar modules, the vehicles, the flags, and maybe even a golf ball. Comparing these items to the other features of the Moon, your conclusion would most surely be that someone had to create them; there had to be an intelligent creator. You then see the bright blue marble above, and travel to Earth, authenticating your premise when you engage its scientists and citizens. We can find another metaphor here: the Moon can be our lives, and Earth the place of knowledge, or heaven.

Design Evidences in the Cosmos: Divine Improbability

The *anthropic principle* involves the fine-tuning of the universe through the improbable balance of many physical constants. Life could not exist on Earth if there was a 1 percent change in sunlight, if the planet were 2–5 percent closer to the sun, or if it were 2–5 percent further from the sun. If the planet Jupiter's distance from Earth were greater, our planet would experience too many asteroid and comet collisions. If Jupiter's distance from Earth were less, Earth's orbit would become unstable. Ross's book *The Creator and the Cosmos* and various websites delineate many factors like these which must prevail together for life to exist on Earth. Ross supports these results with at least 89 reputable scientific references.

Ross estimates the probabilities for attaining the necessary parameters to support life on Earth. There are 75 parameters. The probability of all of them being within the necessary boundaries for life is about 10^{-99} . We have a virtual impossibility of life occurring by accident on Earth. We might say that life on Earth is divinely improbable.

The discussion in this chapter leads me to another faith axiom: *There exists exactly one God.* God is a perfectly good being, personal and transcendent, who created and rules the entire universe. God exists and remains within the universe and the souls of mankind; God is immanent. God is the ultimate model of moral goodness. God is infinity in that we cannot totally conceive of Him in this life.

I believe some people accept the faith axiom *not* to believe in God because they simply do not want to deal with the implications of God in their lives. In *The Great Divorce*, C. S. Lewis writes of a situation where people come out of hell to some kind of land where they can see enough glimpses of heaven that they can still choose it over hell. Some people refuse to choose heaven. In particular, an artist of great fame and power refuses because he does not want to lose the power, fame, and adulation he had on Earth that will continue in hell but not in heaven. For a similar reason, the artist chooses not to accept the existence of God because he will lose the power of his self-centeredness.

I first connected Lewis's ideas with mathematics when I read his article "Transposition"; it was indeed an epiphany, because it connected the notion of higher dimensions in mathematics to the Christian faith. The article was a basis for this chapter. Lewis had a much greater imagination about mathematics than he gave himself credit for. He detested computational arithmetic but loved the reasoning required in geometry. Lewis once said, "I read algebra, devil take it!" He underestimated himself. He may not have cared for the part of mathematics that was computational, but he enjoyed the part involving creative thinking.

In conclusion, I invite my reader's imagination to join mine and assume that you have reached heaven, the $(n +1)$st dimension. What do you think might happen each day (if such an entity exists)? We might get up, have breakfast, and read a newspaper in which the sports section discusses the score of the baseball game pitting the Southern Baptists against the Lutherans. Would both teams win? Then we might sit on a mattress of clouds and drink butterscotch milkshakes, served by angels. Later, we might play golf. But would we score all holes-in-one, or would there be enough of an element of imperfection that our score might vary? Would we go to movies we have not seen? Would we read new books? That might not be necessary if God bestowed all His knowledge upon us. Or, would we just sit on the top of a mountain and explore the wonder?

Although some of these habits might be interesting, I think there is more Glory and mystery of God that He will reveal to us in heaven. This notion is based on the following passage, which takes place on Earth, not in heaven:

> **Exod 33:18-23:** *Moses responded, "Then show me your glorious presence." The Lord replied, "I will make all my goodness pass before you, and I will call out my name, Yahweh, before you. For I will show mercy to anyone I choose, and I will show compassion to anyone I choose. But you may not look directly at my face, for no one may see me and live." The Lord continued, "Look, stand near*

me on this rock. As my glorious presence passes by, I will hide you in the crevice of the rock and cover you with my hand until I have passed by. Then I will remove my hand and let you see me from behind. But my face will not be seen."

But, in heaven, the scripture may have a different meaning. We can see His glorious presence, but it may take the infinitely many days of eternity. The following is my theory, based on the metaphor of higher dimensions. It goes like this. Each day, God might take us one dimension closer to His wisdom and the awesome cosmos He created for us. The first few days might go like the following:

Day 1, $(n + 1)$st dimension: A brief glimpse of heaven, tour the area, find my residence, and be reunited with Elaine.

Day 2: $(n + 2)$nd dimension: Meet my mother, get to know her, and understand her death.

Day 3: $(n + 3)$rd dimension: Meet my other loved ones and friends who have passed away.

Day 4: $(n + 4)$th dimension: First tour of outer space, including Jupiter, Mars, Mercury, and Venus.

Day 5: $(n + 5)$th dimension: First answers to questions from history: What happened on Flight 93 on 9/11/01, and who killed JFK and why?

Day 6: $(n + 6)$th dimension: First mathematics answer day: Explain and then show me a proof of Goldbach's conjecture (considered in a later chapter) and one other unsolved problem.

Day 7: $(n + 7)$th dimension: Second tour of outer space: We travel to the first planet beyond Earth that has life, meet the beings there, and encounter their culture. If there is no life on any other planet, we travel inside the Sun and get a complete explanation of how it works.

Day 8: $(n + 8)$th dimension: See a DVD of the Big Bang and a few days or years thereafter.

Day 9: $(n + 9)$th dimension: Answers to two challenging theological questions: Did somebody make you? Was there time before the Big Bang?
And so it goes.

Let your imagination take you even further. You may wonder if these next dimensions would run out. I don't think so, because heaven is infinite-dimensional. Every day can take us to a next dimension. Wow! I can't wait. Won't you join me?

Chapter 8

Higher Dimensions: Metaphor for the Trinity

Nature shows us only the tail of the lion. But I do not doubt that the lion belongs to it even though he cannot at once reveal himself because of his enormous size.

—Albert Einstein

In the previous chapter, we used a variety of concepts such as photographs, music, and scenes from nature to introduce the notion of higher dimensions. Those notions of higher dimensions provided metaphors for the soul, heaven, and the existence of God. We now use our imagination spatially or geometrically to form an idea of fourth, fifth, sixth, and even higher dimensions. Such a spatial viewpoint will provide a metaphor for understanding the Christian concept of the Trinity: the notion that God, Jesus Christ, and the Holy Spirit are the same entity, as well as the fiery furnace, transfiguration, and Christ's appearance to His disciples after His resurrection.

Historically, mathematicians, physicists, artists, and philosophers virtually insisted on thinking of the fourth dimension in some kind of spatial or mystical context. In 1915, Einstein opened the door to the notion of a fourth dimension when he considered not only length, width, and height but time in his famous theories of relativity. The task of conceiving of the fourth and higher dimensions spatially remains daunting. Nevertheless, in this chapter we will try to imagine dimensions spatially, abandoning the notion of time until the next chapter; but the way we do so is as much a *process* as it is spatial.

Looking at the Dimensions in a Different Way

In the previous chapter, we discussed *higher dimensions* as ways of looking at things such as a person or a beautiful scene, starting with a photo and moving to a real person or to being in the beautiful scene. For purposes of expanding our use of higher dimensions as metaphor, we'll use a geometric perspective and start with a zero dimension.

0D: The Zero Dimension

For the zero dimension (0D), think of the notion of a *point* from geometry. A point has no length, width, or height. Although we cannot see a point, we build the concept by thinking of a period at the end of a sentence, the tip of a pencil, the tip of a thumbtack, a location, or—with some stretch of the imagination—a speck of sand. Mentally, I usually attach at least length and width to the notion of a point, even though those dimensions are not there.

Point   Tip of a pencil   Tip of a thumbtack

Figure 8.1 Ideas of points

1D: The First Dimension

We have lots of ideas in our experience that build the idea of the first dimension.

For example, consider a straight line drawn with a ruler, or a piece of string held tightly, or the edge of a page of this book. For the first dimension (1D), use your imagination to start conceiving a process that will apply to higher dimensions. To conceptualize the idea of a higher dimension, stretch your imagination by thinking of a point as a small cube, such as a die or a grain of sand. Here we are cheating, so to speak, by attaching length, width, and height to an object, but it will help you think conceptually of using such a point to get a line. If I make a small dot with a pencil, it has at least length and width, but also height, which is the extremely thin thickness of the pencil mark.

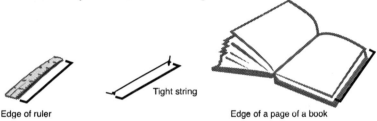

Edge of ruler   Edge of a page of a book

Figure 8.2 Ideas of first dimension

Starting with a point thought of as a single cube, imagine taking an infinite number of points (cubes) and moving them in one direction into the first dimension. You might also think of dipping the cube in paint and dragging it into the first dimension. For the first dimension, this thought is easy to conceive because we have knowledge of the first dimension. The resulting line is a representation of the first dimension. A line has length but no width or height, and it consists of an infinite number of points.

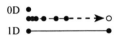

Figure 8.3 0D to 1D: dragging into the first dimension

2D: The Second Dimension

To get an idea of a plane, think of a playing card, the floor of a room, a desktop, or a page of this book.

Width, length, 2D   Card

Table top

Piece of paper

Figure 8.4 Ideas of planes

For the second dimension (2D), imagine taking an infinite number of lines and moving them in one direction in the second dimension, or dipping a line in paint and dragging it into the second dimension. This would give you a plane. A plane has length and width but no height or thickness. This thought is easy to conceive because we have knowledge of the second dimension.

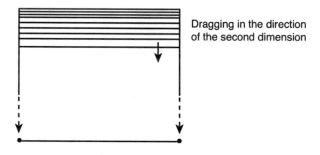

Dragging in the direction of the second dimension

Figure 8.5 1D to 2D: dragging into the second dimension

3D: The Third Dimension

To get an idea of a cube, think of a stack of playing cards, or a ream of paper, or this book lying shut.

Stack of cards   Book

Figure 8.6 Ideas of cubes

For the third dimension (3D), imagine taking an infinite number of planes and stacking them or moving them in one direction in the third dimension, or dipping a plane in paint and dragging it into the third dimension. This thought is easy to conceive because we have knowledge of the third dimension. The result gives us length, width, and height and forms an imagined third dimensional (3D) cube.

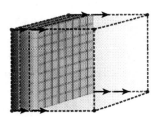

Figure 8.7 2D to 3D: dragging into the third dimension

4D: The Fourth Dimension

Are you ready for the fourth dimension (4D)? The only concept we have of a fourth dimension is that of time, but we ignore that here because we want to examine higher dimensions, but only spatially or geometrically. We have an unfamiliar concept, but we are trying to conceive it using the process we used in creating the first, second, and third dimensions.

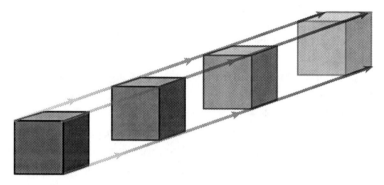

Figure 8.8 3D to 4D: dragging into the fourth dimension. A hyperline (rather than a hypercube as defined in most mathematical literature).

How do we now imagine the fourth dimension? In the previous chapter, we took a 3D object and added music, sound, and an inner soul. But what about geometric objects? Historically, philosophers, artists, scientists, and mathematicians have found it almost impossible to conceive of or imagine the fourth dimension in a spatial or geometric manner. Here we attempt to imagine the fourth dimension spatially, but our thinking is more of a *process* than a picture. Think of the process we used in earlier figures. In the previous section we have an actual cube; we don't have to cheat to think of it as we did with a point as a cube in 0D. To get a concept of a fourth dimension, imagine taking the 3D cube and moving it in the direction of the fourth dimension in a manner analogous to the way we moved the point. We call the 4D object a *hyperline*. In most mathematical literature, it's called a *hypercube* or *tesseract*. We'll call it a hyperline and use the terminology differently here to enhance the process of imagining the fourth and higher dimensions.

The impediment to our imagination in trying to picture the fourth dimension is attempting to draw a 4D figure in 2D. We use a 2D drawing to draw a 3D figure, which is used to imagine a 4D drawing. We have a cube moving in the fourth dimension. You have to use your imagination to envision it.

5D, 6D, 7D: Higher Dimensions

Look again at the earlier figures. Can you stretch your imagination even further now to conceive the fifth, sixth, and seventh dimensions? How about more? When we get a hyperpoint, we move it infinitely to make a hyperline. Then, we move the hyperline infinitely to a make a hyperplane and the hyperplane to make a hyper-hyperpoint. From each new hyper…point, we imagine or visualize the next three dimensions. The situation is analogous to how we name numbers using periods.

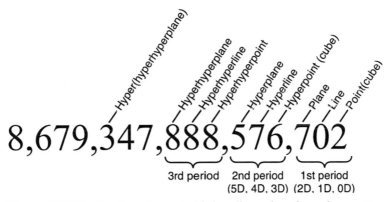

Figure 8.9 Moving from lower to higher dimensions is analogous to how we name numbers using periods.

Graphs, Dimensions, and *n*-tuples

We have considered higher dimensions conceptually and geometrically. Now, we'll look at them more mathematically. We can visualize the first three dimensions as 1-tuples, 2-tuples, or 3-tuples. Rene Descartes (1596–1650), a great mathematician and philosopher, first introduced the idea of attaching an ordered set of numbers to a dimension. Until then, Euclid thought of dimensions only in terms of qualities of shape. Descartes made the process more mental, mathematical, or analytic, but the idea was not compelling to mathematicians until Bernhard Riemann developed his advanced mathematics of differential geometry around 1854. Then, Einstein made the use of four dimensions a practicality in 1915 when he modeled the universe not only with spatial dimensions of length, width, and height, but also with time. Mathematics was ready to take the next stage in the use of higher dimensions by considering the concept of an *n*-tuple.

One Dimension, a_1

A 1-tuple (pronounced "one-tuple") is one real number, a_1, that corresponds to a number on a real number line.

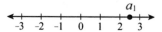

Figure 8.10 A 1-tuple, or number a_1, as seen on a number line

The graph of an equation or inequality in one variable is a set or a picture of all the real numbers, or 1-tuples, that make it true when substituted for the variable. For example, here you see the graph of the inequality $x \le 2$ on a number line. It is a visualization of all the numbers that satisfy the inequality.

Figure 8.11 Graph of inequality $x \leq 2$ in one variable on a number line

Two Dimensions, (a_1, a_2)

A 2-tuple (pronounced "two-tuple") is a sequence of two real numbers, (a_1, a_2), normally called an *ordered pair*. We get a visual picture of ordered pairs by considering a plane with a two-dimensional coordinate system consisting of two lines at right angles. To graph an ordered pair or point in 2-space, we start at the origin (0,0) and move horizontally to the number a_1, the first coordinate. We move right if the number is positive, move left if it is negative, and stay put if it is 0. From that point on the horizontal axis, we move vertically to the number a_2: We move up if it is positive, we move down if it is negative, and we stay put if it is 0. A set of these ordered pairs constitutes the graph of an equation with two variables. If you've ever worked with graphing data in such a coordinate system, you've worked with ordered pairs.

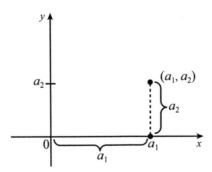

Figure 8.12 An (x,y) two-dimensional coordinate system

Let's graph some ordered pairs using a two-dimensional coordinate system. To graph the ordered pair (–3,5), we start at the origin (0,0) and consider the first coordinate –3. Because it is negative, we move to the left horizontally to the number –3. Then, because the second coordinate, 5, is positive, we move up vertically 5 units. The end result is the graph or plot of the ordered pair (–3,5). If you are unfamiliar with this type of graphing, look at the other points on the graph and think out how they were graphed.

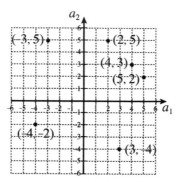

Figure 8.13 Plots of various ordered pairs

The inequality $x \le 2$ has one variable. We graphed it on a number line. The equation $y = x^2 - 1$ has two variables. A solution of this equation is an ordered pair in which each number in the pair corresponds to a letter in the equation. To determine whether a pair is a solution, we substitute the numbers in the pair alphabetically and determine whether the equation is true. For example, the pair (–2,3) is a solution of $y = x^2 - 1$ because $3 = (-2)^2 - 1$ is true. But the pair (1,3) is not a solution because $3 = (1)^2 - 1$ is false.

To graph an equation in two variables, we make a drawing that represents all of its solutions. It is typically graphed on a plane or two-dimensional coordinate system. The solutions of the ordered pairs are plotted, and we often look to see what kind of pattern or direction they take. In this case, the numbers plotted create a parabola. Graphing by hand is a lengthy process for equations, and it is common today to use a computer software package or a hand calculator.

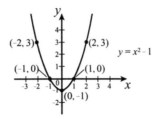

Figure 8.14 Graph of the equation $y = x^2 - 1$

Three Dimensions, (a_1, a_2, a_3)

A 3-tuple is a sequence of three numbers (a_1, a_2, a_3), also called an *ordered triple*. To graph a 3-tuple in 3-space, we need a three-dimensional coordinate system. The axes are placed as shown in the following figure. The line a_3, called the a_3-axis, is placed perpendicular to the $a_1 a_2$-plane at the origin $(0,0,0)$. To help visualize this three-dimensional coordinate system, think of looking into the corner of a room, where the floor is the $a_1 a_2$-plane and the a_3-axis is the intersection of the two walls. To plot or graph a point (a_1, a_2, a_3), we first locate the point (a_1, a_2) in the $a_1 a_2$-plane and move up or down in space according to the value of a_3.

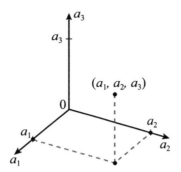

Figure 8.15 A three-dimensional coordinate system

Here are examples of plots of 3-tuples, or points in 3-space, $P_1(2,3,5)$, $P_2(2,-2,-4)$, and $P_3(2,3,0)$.

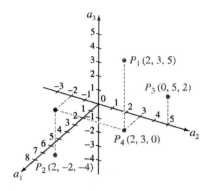

Figure 8.16 Plots of points in a three-dimensional coordinate system

The equation $z = (x^2 + y^2)^{3/2} - 4(x^2 + y^2)$ has three variables. A solution of the equation is a 3-tuple in which each number in the triple corresponds to a letter in the equation. To determine whether a triple is a solution, we substitute the numbers in the triple alphabetically and determine whether the equation is true. To graph an equation in three variables, we make a drawing that represents all of its solutions. The graph of $z = (x^2 + y^2)^{3/2} - 4(x^2 + y^2)$ is shown in the following figure. Creating three-dimensional graphs by hand is extremely difficult, but many computer software packages can create them quickly. Your computer may already have one installed.

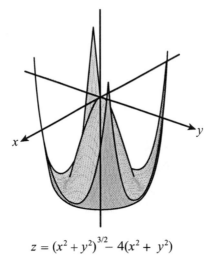

$$z = (x^2 + y^2)^{3/2} - 4(x^2 + y^2)$$

Figure 8.17 A graph of the equation $z = \left(x^2 + y^2\right)^{3/2} - 4\left(x^2 + y^2\right)$

n-Dimensions and *n*-Tuples

We have examined 1-tuples, 2-tuples, and 3-tuples in three dimensions and seen how to graph such entities and related equations. We stretch our mind a bit to draw a graph in three dimensions on a two-dimensional piece of paper. We can use our imagination to try to envision shapes and graphs in 4-space, 5-space, 6-space, and so on, but the task is daunting at best, although we pursued it in the last chapter. But we can consider the fourth, fifth, and sixth dimensions abstractly by considering them in terms of 4-tuples, 5-tuples, and so on. The idea of an *n*-tuple can be abstracted as a mathematical symbol representing *n*-dimensions. To connect geometry with algebra, we regard a point in *n*-space with an ordered sequence or string of real numbers, called an *n*-tuple:

$$(a_1, a_2, a_3, \ldots, a_n)$$

The dimensions come sequentially from real number coordinates

$$a_1, a_2, a_3, \ldots, a_n$$

Projections

When we work backward from a higher dimension to a lower dimension, we have a *projection*. Look carefully at each of the four graphs that follow.

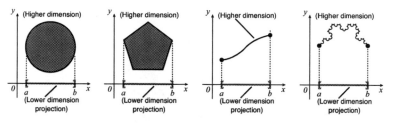

Figure 8.18 Four graphs that have the same projections (prejections) onto the *x*-axis

Each graph is a set of ordered pairs (a_1, a_2). If you eliminate the second coordinate in each case, you get a set of numbers that forms an interval on the *x*-axis. That interval is the projection of the set of ordered pairs in the graph. (I prefer the word *prejection* because it suggests finding what went before. But that is not standard mathematical language.) Recall what we have discussed about lower dimensions being inadequate to describe higher dimensions. How do we interpret that notion with graphs? In each of the graphs, all four lower 1-dimensional projections of the four higher 2-dimensional regions are the same interval $[a, b]$. That interval is shown here.

Figure 8.19 Interval that is the projection of each graph

Suppose we want to find out the shape of the 2D region from examining the 1D interval. The 1D interval is inadequate to tell you much about the 2D shape, other than its width. Infinitely many possible shapes in the second dimension can't be determined from the first dimension.

Let's consider projections from 3D to 2D. Look at the following two graphs.

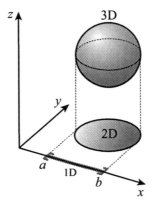

Figure 8.20 A projection from 3D to 2D

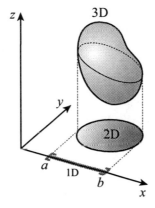

Figure 8.21 A projection from 3D to 2D

The 2D oval shape is the same projection for each of the two 3D graphs. But the 2D shape is inadequate to predict the 3D shape. Using our imagination here and the mathematics of graphing, this discussion lends further support to our assertion that a lower dimension is inadequate to describe a higher dimension.

Flatland People

Another spatial model for imagining the fourth dimension is based on the well-known fictional book *Flatland*, by Edwin A. Abbott, and later extended in the book *Flatterland*, by Ian Stewart. David Neuhouser's book *Open to Reason* and Rudolph Rucker's *The Fourth Dimension: A Guided Tour of the Higher Universes* develop the spatial notions of the fourth dimension further.

Residents of Flatland live in two-dimensional space. Even though objects in 2D have no thickness, imagine that they are thick enough that people can see each other's edge, but that's all. To imagine this, think of cutting geometrical shapes out of paper and floating them on water. But your eyes can only see the edges.

Written in 1884, *Flatland* was a social satire on Victorian society in the 1880s. For example, Abbott writes the following:

Our Women are Straight Lines. Our Soldiers and Lowest Classes of Workmen are Triangles with two equal sides [isosceles triangles]. Our Middle Class consists of Equilateral or Equal-sided triangles. Our Professional Men and Gentlemen are Squares and Five-Sided Figures, or Pentagons. . . . Next above these are the Nobility, of whom there are several degrees, beginning at Six-Sided Figures, or Hexagons, and from thence rising in the number of their sides til they receive the honourable title of Polygonal, or many-sided. Finally when the number of sides becomes so numerous, and the sides themselves so small, that the figure cannot be distinguished from a circle, he is included in the Circular or Priestly order; and this is the highest class of all.

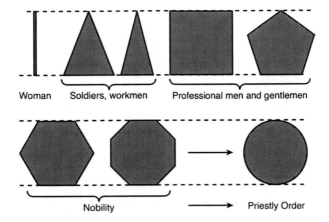

Woman   **Soldiers, workmen**   **Professional men and gentlemen**

Nobility   →   **Priestly Order**

Figure 8.22 People's shapes in Flatland

Note the limiting process that prevails from women to men to priests; a certain sexism existed. People can distinguish themselves by the number of vertices; and women can become virtually invisible by rotating their bodies accordingly. Think of looking toward a needle lying at eye level on a table and rotating an endpoint toward your eye. You can draw your own conclusions about Victorian society based on this visualization.

Think of the frustration a person in Flatland would have, trying to imagine the third dimension. Imagine a square, named A. *Square* in the book, sitting alone in a room. A. *Sphere* in the story miraculously enters A. Square's plane and progressively moves through it. As A. Square watches, he first sees a dot like a pinhead. But it soon widens and looks like familiar straight lines in his world. The line reaches a maximum length, begins shortening back to a dot, and then disappears. This all occurs while A. Sphere in his world notes that he touches A. Square's plane. The dot soon becomes a circle with a widening diameter as the sphere moves further through the plane. The circle reaches a maximum diameter and then begins to decrease until it becomes a point and disappears.

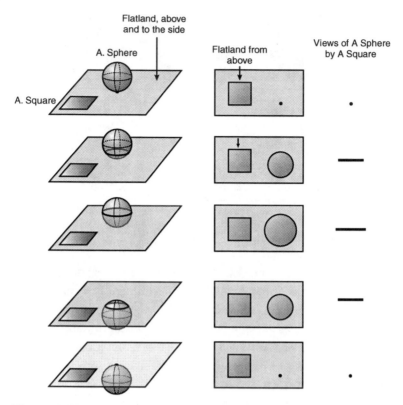

Figure 8.23 A. Square and A. Sphere compare their 2D and 3D worlds

A. Square, in great confusion, wants to understand what is happening, but A. Sphere can't get it across. Finally, in a desperate attempt to explain, A. Sphere elevates A. Square out of his two dimensions and into the third dimension, making him a cube. A. Square, in his mental and physical liberation, soon asks A. Sphere about the fourth dimension. This time, A. Sphere is perplexed. Only A. Square has had the exhilaration of moving from a lower dimension to a higher one. Note how we again find a lower dimension inadequate to explain a higher dimension.

Another Metaphor for the Existence of God

How can we model the existence of God using spatial dimension concepts as a metaphor? I like to think of the Flatland people as Flatwater people. Take a tank of water, cut out some people shapes from a piece of paper, and float them on the water. Then, take a rubber ball or a softball, and touch the water in a small sphere. Now, think of the ball being symbolic of God; the flat plane of the top of the water is our 3D universe, even though it is 2D in the illustration. Imagine the small touching sphere (point) as a union of all the evidences of God that we have explored. It's not much different than the sphere and square example described in the previous section. Oversimplified as this example may seem, it is an effective illustration of how little of God we know but how magnificent He is. The metaphor comes from the impossibility of imagining all of God (the ball) from just the hint (the point) we have of where it touches the surface of the water where we exist.

Solid
sphere

Figure 8.24 Flatwater people as their world is invaded by a solid ball

To enhance the metaphor, let's take the illustration even further. We change the ball to a 3D graph of the equation of the elliptic paraboloid $z = (x - 3)^2 + (y - 4)^2$. Think of its inner volume, which is now infinite, as the symbol of God; the xy-plane as our universe, still similar to our living on the surface of the water; and the point of tangency, in the xy-plane of the surface of the water, as the part of God we experience in our lifetime. We used the ideas in the circle in the preceding chapter as clues to the existence of God. Here, we see how that experience is only a hint—a small dot or circle. The volume of the shape is infinite, as is God. He touches us only minutely at the point. There is much of Him that we do not know.

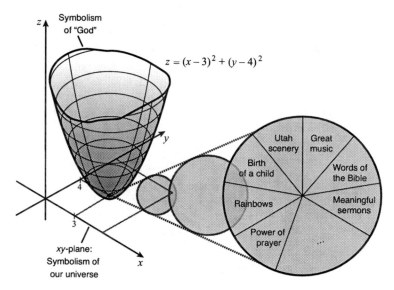

Fig 8.25 The equation $z = (x - 3)^2 + (y + 4)^2$ and its inner volume as a metaphor for God

We can see this dot in our 3D existence much like the circle of evidence in the preceding chapter. There are many ways to debate the existence of God besides those presented in this book. In the final analysis, humankind must make a faith axiom of God's existence. Over the years, as I've pondered the connections between mathematics, the dimensions, and God, I've arrived at a faith axiom: *There exists exactly one God.* God is a perfectly good being, personal and transcendent, who created and rules the entire universe. God exists and remains in the universe and the souls of mankind. God is immanent. God is the ultimate model of moral goodness. God is infinity in that we cannot totally conceive Him in this life.

Can we ever see God completely? Several quotes support the notion that God is infinite and can never be seen completely. In the book *The Works of Jonathan Edwards*, the author says, "Were God to disclose but a little of that which is seen by saints and angels in heaven [the dot at the base of our figure], our frail natures would sink under it.... Such a bubble [us] is too weak to bear a weight so vast. Alas! No wonder therefore it is said, No man can see God and live."

We know the Bible supports this idea as well:

> **Exod 33:19-20:** *The LORD replied, "I will make all my goodness pass before you, and I will call out my name, 'the LORD,' to you. I will show kindness to anyone I choose, and I will show mercy to anyone I choose. But you may not look directly at my face, for no one may see me and live."*

> **John 1:18:** *No one has ever seen God. But his only Son, who is himself God, is near to the Father's heart; he has told us about Him.*

> **1 Tim 6:16:** *He alone can never die, and he lives in light so brilliant that no human can approach him. No one has ever seen him, nor ever will. To him be honor and power forever. Amen.*

Thus, even in another dimension, such as heaven, we may still never fully comprehend God. Is it then so unreasonable that we have difficulty conceiving of Him in our present dimension?

A Metaphor for the Trinity

One of the most questioned and debated tenets of the Christian faith is the concept of the Trinity, which states that there are three facets of God. Although God is one entity, we can distinguish Him in three forms. He is at least the *union* of three distinct persons: the *Father*, the *Son*, and the *Holy Spirit* or *Holy Ghost*.

Author Frederick Buechner describes the Trinity this way in his book *Wishful Thinking: A Seeker's ABC:*

> *(1) The Father:* The mystery *beyond us* or the idea or essence of all reality from which everything exists and flows.

> *(2) The Son (Jesus Christ):* The mystery *among us* or the perfect expression of the essence of reality.

> *(3) The Holy Spirit:* The mystery *within us* or the recognition of the essence of reality inside human beings. It is the belief of Christians that the Holy Spirit comes within them after coming to faith in Christ. In this way, God continues His work by working within us.

Imagine a tuning fork with three tines or prongs as part of an entity we think of as God. The tines represent the Father, the Son, and the Holy Spirit. M. Scott Peck calls the three-pronged tuning fork in these figures a *triadox*, because it is a natural extension of an actual tuning fork, which we have used to describe a paradox. Over the years, the notion of the Trinity has been a subject of great deliberation among theologians, so much so that it could be called a *triadox*, that harmonizes.

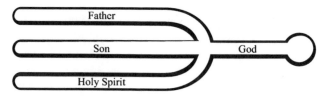

Figure 8.26 The Trinity as a three-pronged tuning fork

We create a metaphor for the Trinity by returning to our Flatwater people floating on top of the tank in order to visualize how the Trinity interacts with their 2D world. As the tuning fork enters the water, only three circles are seen—this represents the Trinity. As the fork is immersed into the water, the circles converge and become not three sources but one as the solid part between the prongs and the handle touches the water. It's as if the Trinity (the three) represent God (the one) that we interact with in this world as the prongs touch the water. And just as the Flatwater people are unable to conceive of the three-pronged tuning fork when the prongs first touch the water, we are unable to conceive of the wholeness and completeness of God and the Trinity when it touches us, even though we live in a 3D world.

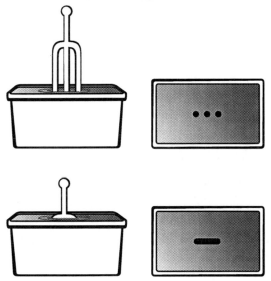

Figure 8.27 The Trinity visualized by the Flatwater people

The notion of the Trinity is difficult enough to comprehend, by Christians and non-Christians alike; but Christians can also experience difficulty in conceiving the entire concept. By allowing ourselves to embrace the metaphor of higher dimensions, those who are inclined toward mathematics may arrive at a more tranquil acceptance of the concept of Trinity. The three aspects of the Trinity become one in a higher dimension.

The Fourth Dimension and Other Spiritual Implications

As mentioned earlier, when we work backward from a higher dimension to a lower dimension, we have a projection. For example, your shadow is a projection of yourself: a 3D-to-2D projection. A piece of music played only on a piano can be thought of as a projection of an orchestral rendition of the same piece. Our three-dimensional existence is inadequate to understand the fourth dimension. We might imagine projections from 4D to 3D to be what seem like miracles in 3D. In his book *Hyperspace*, Kaku says, "…a fourth dimensional being would have almost God-like powers." A person might:

- Walk through walls
- Walk right through mountains, rather than walk over or around them
- Reach their hand through the refrigerator for food, rather than have to open the door
- Disappear or reappear at will
- See through objects with x-ray eyes
- Do surgeries, if they were a surgeon, without cutting the skin

Keep in mind that these are imagined ideas. We would have to experience the fourth dimension to verify that these miracles in the third dimension are projections from the fourth. But do they give you another perspective on God? Many instances from the Bible can be better understood by a fourth-dimensional interpretation.

The Fiery Furnace

If you recall the story from the third chapter of Daniel, his friends Shadrach, Meshach, and Abednego were cast into a fiery furnace by the King of Babylon, Nebuchnezzar. As the Bible then says,

Dan 3:23-25: *So Shadrach, Meshach, and Abednego, securely tied, fell down into the roaring flames. But suddenly, as he was watching, Nebuchadnezzar jumped up in amazement and exclaimed to his advisers, "Didn't we tie up three men and throw them into the furnace?" "Yes," they said, "we did indeed, Your Majesty." "Look!" Nebuchadnezzar shouted. "I see four men, unbound, walking around in the fire. They aren't even hurt by the flames! And the fourth looks like a divine being!"*

This appearance of a fourth person in the fire could have been the shadow or projection of an angel or of God appearing from the fourth or higher dimension.

The Transfiguration

Another way we can interpret miracles as projections of the fourth dimension is with the *transfiguration*, an event described in Matthew 17. Christ takes three of his disciples up a high mountain:

Matt 17:2-3: *As the men watched, Jesus' appearance changed so that his face shone like the sun, and his clothing became dazzling white. Suddenly, Moses and Elijah appeared and began talking to Jesus.*

Moses died in 1406 BC, and the transfiguration was after 28 AD. Elijah was John the Baptist, who was killed after 26 AD. Both of them reappeared in this situation and were talking to Jesus. Soon,

Matt 17:5: *...a bright cloud came over them, and a voice from the cloud said, "This is my beloved Son, and I am fully pleased with Him. Listen to him."*

This event was a vision of Christ's divinity and God's affirmation of what Christ had done and was to do. This event could have been the shadow or projection of Moses, Elijah, and God appearing from a fourth or higher dimension.

Christ's Appearance to His Disciples

The last event we might consider as a projection of the fourth dimension on the third involves the numerous reappearances of Christ after the Resurrection, described in John 20. Christ's shadow or projection may have appeared through the walls as a projection from the fourth dimension. Here is a related Biblical quote:

> **John 20:26-27:** *Eight days later the disciples were together again, and this time Thomas was with them. The doors were locked; but suddenly, as before, Jesus was standing among them. He said, "Peace be with you." Then he said to Thomas, "Put your finger here and see my hands. Put your hand into the wound in my side. Don't be faithless any longer. Believe!"*

Conclusion

In this chapter, we have introduced *n*-tuples, graphs, and geometric interpretation as mathematical models for a more refined understanding of how a lower dimension is inadequate to describe a higher dimension. In particular, our third dimension is inadequate to describe the fourth dimension. We have no way of knowing in our 3D world whether the Trinity can be considered a projection of a Godly fourth, or higher dimension, or whether incidents like the fiery furnace, the transfiguration, and the appearance of Christ happened as projections of entities in the fourth dimension. But knowing that our 3D perception is inadequate to understand 4D provides a kind of model for deeper understanding of the mystery. In the ways of mathematics, our thinking provides greater understanding of God.

Chapter 9

Higher Dimensions: String Theory, Relativity, and the End Times

---

We live in a materialist culture in which the denial of spirit is sanctioned. Instead of employing our science to look more deeply for God, we use it to declare Him out of bounds. In some universities it is actually fashionable to teach what cannot be measured does not exist. The arrogance of it is extraordinary. We presume that our little rules and clocks can encompass Reality. Rather than recognize our limitations, we strive to limit the world. Unable to cut it down to our size, we deny the immeasurable. In some ways our arrogance masks our fear. We are afraid of the immeasurable Spirit, the power of love beyond all comprehension. We want to be in control of life; God forbid that life should be in control of us.

—M. Scott Peck

Greek philosophers Plato and Aristotle were credited with the courtship and *marriage*, or integration, of science and religion. The more science they studied, the more convinced they became that the heavens and the earth certified the existence of God. The solar system was like a finite, self-contained structure in which God seemed understandable. Tycho Brahe (1546–1601) was a Danish Astronomer credited with the most accurate astronomical observations of his time. His data were used by his assistant Johannes Kepler (1571–1630) to derive his famous laws of planetary motion. In 1572,

Brahe discovered a new star, actually a supernova, in the constellation Cassiopeia. A *supernova*, one of the most energetic explosive events known in astronomy, occurs at the end of a star's lifetime, when its nuclear fuel is exhausted and it is no longer supported by the release of nuclear energy. (The middle "star" in the belt of the constellation Orion is a supernova.) Five years later, Brahe observed a new comet in the heavens. These discoveries began the divorce of science and religion.

Subsequent discoveries by Kepler that the orbits of planets were elliptical, discoveries by Copernicus and Galileo that the planets revolve around the sun, and other findings by Brahe, crumbled the marriage of science and religion even further. Paradoxically, the marriage was dissolved by Sir Isaac Newton (1642–1727), a man as occupied with his Christianity as with science and mathematics, when he developed his Universal Law of Gravity; it established that all particles of matter in the universe feel a force of attraction between each other, not just from the Earth.

The growth of scientific knowledge flourished; but in the process, humans—carried away with newfound intellectual power—began to conjure the notion that they could figure it all out by themselves and no longer needed a concept of God. In effect, science became a god unto itself. Instead of pursuing God, man pursued science; science became its own false idol—a false infinite, so to speak.

Dallas Willard, in his book *Hearing God,* says the following:

The discoveries of the immensity of space and of the forces of nature—which appear to determine everything that happens and seem to run their course with no assistance from the hand of a personal God—can be quite overwhelming. When the great French mathematician and astronomer Pierre Simon de Laplace [1749-1827] presented Emperor Napoleon with a copy of his book on celestial mechanics, the emperor asked him where God, a supernatural being beyond the natural world, fit into his system. Laplace indignantly drew himself up and replied, "Sir, I have no need of any such hypothesis!" According to the current model of natural sciences, nature proceeds without invoking God.... [T]he university system stands in world culture as the source of unquestioned authority so far as knowledge is concerned.... [I]t currently throws its weight behind a picture of reality without God, a picture in which human beings are entirely on their own. Regardless of what the recognized system of education might say of itself for public relations purposes, it presumes in its processes that you can have the best education possible and be ignorant of God.

In his recent book *Rebuilding the Matrix: Science and Faith in the 21st Century*, Denis Alexander writes, "the scientific enterprise is full of experts on specialist areas but woefully short of people with a unified worldview." I have a Ph. D. in Mathematics Education. The point is not that I have such a degree, but that Ph. D. means *Doctor of Philosophy*, which reflects the age when science was integrated with religion or philosophy. Yet because of the divorce of science and religion, I never took a philosophy course in my graduate work. I missed out on a way to integrate my faith with my profession.

Willard asserts further, in frustration,

Today it is simply assumed that scientific knowledge excludes the presence of God from the material universe of which we human beings are supposed to be a pitifully small and insignificant part. This is called "naturalism." Deism is the belief held by a person who does not embrace any particular religious faith, but does believe in God, a God who for the most part created the world and left it to its own devices.

Theism, according to Alexander, is

...the view held by the Keplers, Galileos, and Boyles of the 17th century that God not only created the universe in the beginning, but is in a moment by moment relationship with it, actively upholding and sustaining both its existence and its properties. In "theism" the universe is viewed as contingent ("dependent") upon God's continuing creative actions, whereas in "deism" the universe is noncontingent ("independent"). Science and mathematics can go beyond religious, cultural, and political differences to explore the universe and all its workings, but it cannot attach meaning and purpose to such knowledge. But, the faith of a theistic scientist, or mathematician, can attach meaning and purpose which transcends this knowledge.

Alexander concludes his book by saying

...theism provides a unified worldview that does a remarkably effective job in providing a matrix for science in which the validity of scientific knowledge is justified and in which the fruits of scientific discoveries are channeled in ways that affirm human value, justice and care for the environment.

Today there should be a remarriage between science and religion, encouraging scientists and Christians alike to see how sciences such as mathematics are at work in the world around them. For example, as we explore how math and science create metaphors for our faith, we delve more deeply into mathematics: higher dimensions integrated with science, relativity, quantum physics, string theory, and also the philosophy of paradox. The end of this chapter brings the integration of science (relativity and quantum physics), mathematics (higher dimensions), and religion (the Christian view of the end times as depicted in Revelation 21) to fruition.

We consider several concepts of mathematics and science as they apply to Christianity and the Bible. We'll first briefly relate Einstein's Theory of Relativity to higher dimensions. In this context, we consider higher dimensions more scientifically with time as a fourth dimension. Our second goal is to consider a recently developed hot topic from physics called *string theory*. String theory extends our thinking about dimensions from the 4 dimensions of length, width, height, and time to 10 dimensions, which occurred at the beginning of time as we know it: the Big Bang.

The third goal is to relate string theory to the notion of paradox, which is a cornerstone of this book. The fourth goal is to extend the first 10 dimensions of string theory to even higher dimensions and apply them as metaphor to theological issues such as foreknowledge, prayer, pain and evil, and creativity. This extension again uses higher dimensions as metaphor to grasp these Godly concepts. Our final goal is to use the preceding information as puzzle pieces to create an original scientific and mathematical insight into the end times of Revelation 21.

Taking the Leap to an Integration of Science and Religion

I find it interesting, if not appalling, how often I read a book or see a movie in some nonreligious area, only to have it take the reader right to the brink of a spiritual connection—and then stop, fearing to dive into a situation where the subject matter might get integrated into the Christian faith. I understand and respect that such an author may not be a Christian; but when I anticipate a connection to the faith that would enhance the meaning and purpose of the movie, I beg for its inclusion. Such an inclusion of faith would have enhanced films such as *It's a Wonderful Life, Braveheart, High Noon, The Magnificent Seven, Rocky, Cinderella Man, Legends of the Fall,* and *Ice Castles.*

Some years ago, I read the best-selling psychology book *Emotional Intelligence,* by Daniel Goleman. I found the book worthwhile and full of insight as it made a case for all professions to improve their emotional intelligence (EQ) or people skills. As an educator, the book struck in me the chord that all educators would be better at their teaching if they improved their people skills. I felt this so strongly that I gave talks at math conventions based on the idea that students will not care about what we know as teachers until they know we care about them as people. Goleman describes many ways to improve EQ but stops short of discussing a highly effective, integrative, result in the Christian faith—what the Bible calls the "fruits of the spirit":

> **Gal 5:22:** *But when the Holy Spirit controls our lives, he will produce this kind of fruit in us: love, joy, peace, patience, kindness, goodness, faithfulness, gentleness, and self-control. Here there is no conflict with the law.*

Do you want to improve your people skills? Then become a Christian. The Holy Spirit directs the lives of Christians with a bias to action, and those actions convey the fruits of the spirit as listed. I can't imagine anyone not benefiting by enriching their lives with these people skills. And by integrating them with Goleman's idea of emotional intelligence, a person could find a level of fulfillment and success in a Christian life that they may have missed before.

Higher Dimensions, Time, and Relativity

Here, we extend our ideas about higher dimensions considered as ordered n-tuples and see how Einstein's Theory of Relativity affects even math laws as trivial as the Pythagorean Theorem. The Pythagorean Theorem says that the sum of the squares of the sides of a right triangle is equal to the square of the hypotenuse. That is, the hypotenuse is the principal square root of the sums of the squares of the sides.

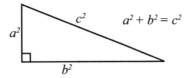

Figure 9.1 The Pythagorean Theorem: In a right triangle, $a^2 + b^2 = c^2$

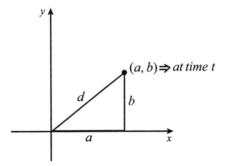

Figure 9.2 The Pythagorean Theorem applied to ordered pairs

Applying the Pythagorean Theorem in a two-dimensional coordinate system, the distance d from the origin $(0,0)$ to (a,b) is $d = \sqrt{a^2 + b^2}$. Albert Einstein established time as the fourth dimension in his Special Theory of Relativity in 1905, which he extended in his General Theory of Relativity in 1917. According to Einstein, we must consider *events* as the true realities—an event is something that happens in space (a,b,c) at a given time t. We can think of time as a fourth dimension, and we can consider ordered 4-tuples, or quadruples (a,b,c,t): *length, width, height,* and *time.*

Although space prevents delving into the details of Einstein's theories, it is of interest to point out that when time is used as a fourth dimension, the Pythagorean Theorem of algebra is changed via Einstein's General Theory of Relativity. Let L = the speed of light. Then, in two dimensions, the distance d from $(0,0)$ to (a,b) is changed from $d = \sqrt{a^2 + b^2}$ to $d = \sqrt{a^2 + b^2 - L^2 t^2}$. Similarly, in three dimensions, the distance d from $(0,0,0)$ to (a,b,c) is changed from $d = \sqrt{a^2 + b^2 + c^2}$ to $d = \sqrt{a^2 + b^2 + c^2 - L^2 t^2}$.

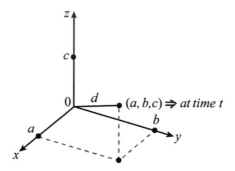

Figure 9.3 The Pythagorean Theorem applied to ordered triples

The Big Bang

Most scientists agree that there was an instant 10 to 15 billion years ago (the precise amount of time is admittedly debatable) when the universe burst into existence in a primordial explosion. Some call it a *dynamic first moment* or a *singularity*, but I will call it by the more common term, *Big Bang* (BB). During this creation event, the universe erupted from an infinitely (or nearly infinitely) dense state and has been expanding outward ever since at increasing acceleration. The BB was not so much an explosion in space as it was the initial rapid expansion of space itself, much like a balloon being suddenly created and expanding.

To understand the concept of an infinitely dense state, think of a bowling ball, and the formula for density:

Density = Mass/Volume

$$D = \frac{m}{V}$$

Imagine a machine that can compress the bowling ball to the size of a baseball. The mass is the same, but the volume is smaller, so the density is higher. Then, find a supermachine that compresses the baseball-sized object down to the size of a jelly bean. Again, the mass is the same, but the volume is even smaller, so the density is even higher. Next, envision compressing the jelly bean down to the size of a grain of sand. Continuing the process indefinitely, we approach as a limit an object of infinite density. If we really stretch our imagination and envision the entire universe compressed down to the size of a grain of sand, and then to a minuscule-sized particle smaller than an atom, and we are closer to conceptualizing the infinitely dense state at the Big Bang.

Some scientists hypothesize that the initial singular entity had a volume of about 10^{-100} cc, infinitely smaller than a grain of sand or an atom. The heat level at this time was roughly F—roughly 2 quadrillion quadrillion degrees Fahrenheit—much hotter than a firecracker or a nuclear device. This is more heat and energy than humankind could generate with all of its many scientific advances.

It is impossible for scientists to validate the Big Bang empirically—that is, by repeating an experiment over and over. Nevertheless, science has pieced together enough evidence of other kinds to approach the conclusion.

In mathematical logic, it is a known that if the statement

If P, then Q

is true, it cannot necessarily be concluded that the converse

If Q, then P

is true. Supposing the Big Bang occurred, then many facts can be deduced by physicists. Michio Kaku lists four of them in his book *Hyperspace*:

Q_1: The stars are receding from us at fantastic velocities.

Q_2: The distribution of the chemical elements in our galaxy is in almost exact agreement with the prediction of heavy-element production in the BB and in the stars.

Q_3: The earliest objects in the universe date back 10–15 billion years, in agreement with the rough estimate for the BB. There is no evidence for objects older than the Big Bang.

Q_4: The BB produced a cosmic "echo" reverberating throughout the universe that should be measurable by our instruments. This radiation is called *blackbody radiation.*

It is not our goal to develop such physics, just to point out that scientists have established a sequence of true statements like

If BB, then Q_1

If BB, then Q_2

If BB, then Q_3

If BB, then Q_4

⋮

They can't assert that the converse of each statement is true in order to deduce there was a BB. But, as scientists prove more of these "If P, then Q" statements, they become more assured of their conclusion about the BB.

Few scientists today, Christian or otherwise, deny the occurrence of the BB. The debate centers on who, if anyone, enacted the BB. Did it come out of some kind of primordial soup, or did God instigate it? My faith axiom is that *God instigated the Big Bang.*

Was there time before the BB? Did God cause the BB to happen, thus creating the universe as we know it today? Exploring the issue of time in depth would take us far awry into the depths of science and the theory of relativity. Entire books have been written on the subject; among them *A Brief History of Time* and *The Universe in a Nutshell* by Stephen Hawking, and *Time and Eternity: Exploring God's Relationship to Time,* by William Lane Craig. Hawking, in *The Universe in a Nutshell,* comments that he and Roger Penrose "proved that in the mathematical world of general relativity, time must have a beginning in what is called a big bang." Craig makes a case for the fact that God created time but "before" the Big Bang there was no "time." God was so infinitely dimensional that the only reason God invented time was to accommodate us finite human beings.

Picture time on the following interval:

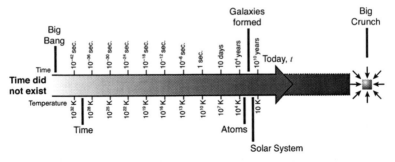

Figure 9.4 Time from the Big Bang to the Big Crunch

It is intriguing to note that although mathematicians can create a set of numbers by subtracting 1 year, 2 years, and so on from the time of the Big Bang, T, as follows,

$$\left\{ ..,T-5,T-4,T-3,T-2,T-1,T \right\}$$

in principle time did not exist until T. This stretches our human understanding. It is my faith axiom that *God is the creator of time and that God exists independent of time*. I base this premise on the writings of Hawking, Craig, and the first verse of the Bible:

> **Gen 1:1-2:** *In the beginning God created the heavens and the earth. The earth was empty, a formless mass cloaked in darkness. And the Spirit of God was hovering over its surface.*

Higher Dimensions and String Theory

It is said that "Mathematics is the Queen of the Sciences," because most scientific theories and research either use or express themselves using mathematics. We are about to touch on some uses of mathematics in science and theology. We examined the idea of an n-tuple

$$\left(a_1, a_2, a_3, a_4, ..., a_n \right)$$

geometrically. Now, let's look at n-tuples scientifically using higher dimensions.

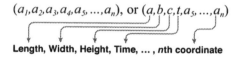

$$(a_1, a_2, a_3, a_4, a_5, ..., a_n), \text{ or } (a, b, c, t, a_5, ..., a_n)$$

Length, Width, Height, Time, ... , nth coordinate

Figure 9.5 An n-tuple in terms of physics concepts of length, width, height, and time

Einstein considered four dimensions (a, b, c, t): length, width, height, and time. Recently, science has hypothesized meanings for the fifth through the tenth dimensions. Insight into those dimensions comes from *superstring theory* or what it usually called *string theory*. To grasp string theory, we start at the Big Bang. The theory asserts that in the first 10^{-43} of a second after the BB, there were six more dimensions or strings that were active and expanding in addition to length, width, height, and time.

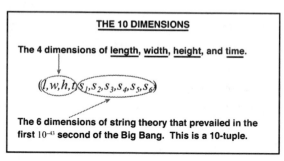

THE 10 DIMENSIONS

The 4 dimensions of <u>length</u>, <u>width</u>, <u>height</u>, and <u>time</u>.

$$(l, w, h, t, s_1, s_2, s_3, s_4, s_5, s_6)$$

The 6 dimensions of string theory that prevailed in the first 10^{-43} second of the Big Bang. This is a 10-tuple.

Figure 9.6 The 10 dimensions in the first 10^{-43} of a second after the Big Bang

These dimensions or strings are the basic composition of matter. How can we visualize this? Look at the following figure. Brian Greene, in his book *The Elegant Universe,* helps us explore how scientists view smaller and smaller entities of matter. Let's start with an object we can grasp, such as a golf ball. Scientists first discover that the golf ball is made of molecules, then they find that molecules are made from atoms, and that atoms consist of electrons, protons, and neutrons, which are made up of smaller entities called quarks. Quarks and electrons are then made up of strings. In a sense, we are examining dimensions by looking down an infinitely descending staircase. Who knows what might come next? In a more recent book, *The Fabric of the Cosmos*, Greene adds, "According to superstring theory, every particle is composed of a tiny filament of energy, some a hundred billion billion times smaller than a single atomic nucleus." And Kaku, in his book *Hyperspace,* describes strings by saying, "according to this theory, matter is nothing but the harmonies created by this vibrating string. Because there are an infinite number of harmonies that can be composed for the violin, there are an infinite number of forms of matter that can be formed out of vibrating strings."

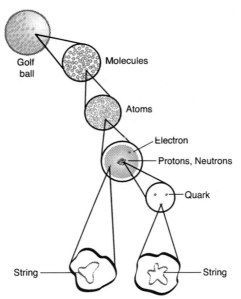

Figure 9.7 The breakdown of matter: molecules to strings

According to Hugh Ross,

Strings are less like strings than they are vibrating, rotating elastic bands. They are greatly stretched at the extremely high temperatures of the first split second of the universe's existence. At the lower temperatures since then, they are contracted to such a degree that they behave like points. String theories do not work in three space dimensions. They need much more room to operate. However, they need that room—six extra dimensions—only for a moment, just a split second after the initial creative burst. From that moment on, these six extra dimensions are no longer necessary to the universe's development. So what happened to the six?

Ross answers by saying, "To this day, the six other dimensions remain curled up everywhere, at every location within our four still-expanding dimensions of length, width, height, and time." The situation is like taking a two-dimensional piece of paper and curling it up tightly on an axis through two corners. If we stand at a distance and look at the twisted paper, it appears to be a one-dimensional line.

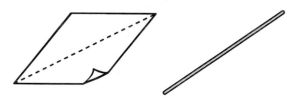

Figure 9.8 A two-dimensional piece of paper rolled up into what then appears to be one dimension

At the extreme temperatures of the first 10^{-43} seconds of the BB, certain kinds of particles called *quarks* reacted in vibration to stay separate from each other. As the temperature cooled after that 10^{-43} seconds, the vibration ceased, and the quarks joined together to form larger particles called *photons* and *electrons*. After this joining or curling up, four dimensions were sufficient to describe nature as we know it.

Paradox and String Theory

A point of comparison between string theory and psycho-spiritual growth is worth noting here. A vibration is a back-and-forth motion or change between two opposing locations in a regularly repeated motion or dynamic resonance. Examples are the beat of a heart, the vibration of sound, the swing of a pendulum, radio signals, and the change with time of the electric and magnetic fields in an electromagnetic wave. In each example, we think of the vibration back and forth as one simultaneous entity. Strings, the foundation of matter, work together in a dynamic tension to create quarks and electrons and are at the foundation of our biological being.

A *paradox* is a seemingly contradictory statement that nonetheless may be true. According to Richard Hansen in an article in *Christianity Today,*

Certain kinds of paradox work in harmony analogous to the vibrations of a tuning fork. A tuning fork delivers a true pitch by two tines vibrating together. Muffle either side, even a little, and the note disappears. Neither tine individually produces the sweet, pure note. Only when both tines vibrate is the correct pitch heard. Despite our biases toward one tine or the other, neither side of the paradox should be muffled, even a little. Paradox beckons us into mystery, and offers a wholesome reminder that God is infinitely greater than our ideas about God. Vibrations are a form of harmonious paradox, declaring their truth when two sides of the paradox vibrate in unison.

According to M. Scott Peck in *Golf and the Spirit,*

To think paradoxically, we must hold two opposites in our minds simultaneously. I'm not sure we humans are capable of thinking about two things at once and certainly not about opposites. What we can do, however, is bounce back and forth—vibrate—between opposing concepts so rapidly as to make our consideration of them virtually simultaneous. When we learn how to do so, the two will become One and the opposites a Whole.... If you think you have discovered a great truth, and it is not a paradox, then I suspect you may be deceiving yourself.

The road to trust in a faith axiom can be derived from the resolution of a dynamic tension between *belief* and *questioning*. In that way we are one step further down the road to psycho-spiritual growth.

Note the comparison. Strings, rapid vibrations between opposing entities, are at the foundation of our biological being. At the apex of our being is our psycho-spiritual growth, and it come from a different kind of thinking about paradox. A form of paradox is at both ends of the spectrum of life.

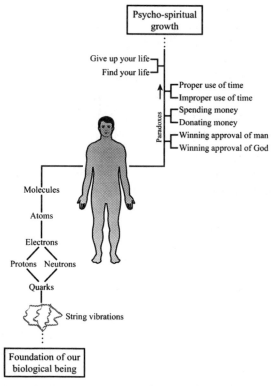

Figure 9.9 Strings, as vibrations, at the foundation of our biological being; paradoxes, as vibrations, at the apex of our spiritual being

Two Theories of Physics Come Together

We look now to paradox in physics. Einstein's theory of gravity has to do with the massive and the very large—planets, stars, and galaxies, for example. The theory of quantum mechanics has to do with the light and the very small—tiny quarks and protons, for example. The Big Bang seems like a paradox. To examine the BB is to examine an entity that is simultaneously very heavy (big, in the sense of mass) and very small, size-wise. The beauty of string theory is it resolves what was perceived by physicists to be a paradox of theories. Two theories, Einstein's general theory of gravity and the theory of quantum mechanics come together in a unified theory by considering 10 higher dimensions—to make sense of string theory, physicists need both theories. String theory resolves the paradox, allowing both theories to be extracted from itself.

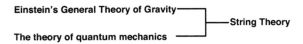

Figure 9.10 String theory is the resolution of the paradox of Einstein's General Theory of Gravity and the theory of quantum mechanics.

Physicists discovered string theory almost by luck, but the mathematics used to describe it lags far behind. Kaku says,

---

Although the theory is well defined mathematically, no one has been able to solve the theory. No one. ... For a string theorist, the fault lies not in the theory, but in the mathematics used to describe it.... [W]e are at a total loss to explain why it works. String field theory exists, but it taunts us because we are not smart enough to solve it. The problem is that while twenty-first century physics fell accidentally into the twentieth century, twenty-first century mathematics hasn't been invented yet.

Typically, in the history of science, the developments of mathematics come way before the needs of science. The mathematics is usually ready and waiting for scientists when their research creates the need for the relevant math. I am frequently asked what mathematicians do. I respond that they are working on the frontiers of mathematical knowledge, and eventually science catches up and needs the mathematics. To know what recent math research is good for, we have to wait for science to catch up. The situation is somewhat comparable to the development of *velcro*. It was invented for use in weightless environments on NASA spacecraft; no use in the normal dimension on earth seemed of value. But humankind caught up, eventually discovering needs for it. Today, we use velcro everywhere as a convenient way to attach things together without snaps, buttons, or zippers. Such was not the case with string theory; physics came first and developed it, but the mathematics necessary to solve it is still lagging behind.

The secret of the mathematics *may* lie in the work begun by a brilliant mathematician from India, Srinivasa Ramanujan (1887-1920), and his study of modular functions. Unfortunately, Ramanujan's work was cut short by his early death at the age of 33 from tuberculosis. Physicists are convinced that the appropriate number of dimensions used in nature is 10 but are at a loss mathematically to adequately explain it. If present-day mathematicians extended the work begun by Ramanujan, then the physicists might be able to unravel the secret.

Another difficulty is trying to test the hypotheses of string theory. According to Kaku, "The theory predicts that the unification of all the forces occurs at the Planck energy, or 10^{19} billion electron volts, which is about 1 quadrillion times larger than energies currently available in our accelerators." In effect, the experiment necessary is comparable to reversing the BB, but the amount of heat and energy necessary to do that is beyond humankind's capabilities. It is my conviction that only God can create such heat and energy. As this narrative progresses, we will consider a theory about how this might happen at the end times.

God in the Higher Dimensions: Theological Issues

We have introduced string theory results in physics regarding 10 dimensions and connected them to the Big Bang. What follows is my theory, speculative and conjectural, on the use of the mathematical framework of higher dimensions to connect higher dimensions, meta-phorically, to the issues of foreknowledge, prayer, pain and evil, and creativity. Let's look again at the 10-tuple of higher dimensions.

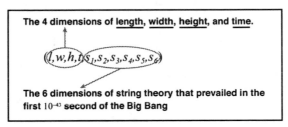

The 4 dimensions of length, width, height, and time.

$$(l, w, h, t, s_1, s_2, s_3, s_4, s_5, s_6)$$

The 6 dimensions of string theory that prevailed in the first 10^{-43} second of the Big Bang

Figure 9.11 Four basic dimensions of length, width, height, and time, plus the six string-theory dimensions

You are looking at the four dimensions of length, width, height, and time and the six dimensions of string theory that prevailed in the first 10^{-43} seconds of the Big Bang. This is a 10-tuple. I arrive at some other faith axioms: *God is the creator of the universe. God is in all the ten dimensions. Many philosophical questions are answered by assuming that God is in yet higher dimensions.*

Let's consider four questions. Keep in mind that the answers come from my theory. The answers, simplistic as they may seem, allow the mathematical framework of higher dimensions to connect higher dimensions, metaphorically, to create a fresh viewpoint on some con-troversial theological issues.

Question 1: Foreknowledge—Does God Have Foreknowledge of Everything That Will Happen?

Humankind has long debated the issue of whether God knows what will happen in the future. Just our study of the probability of prophecy should be enough to verify such foreknowledge on God's part. But if we can think of a higher dimension to interpret situations like Daniel seeing a fourth person in a fire when only three were placed there, or Christ reappearing after the Resurrection and walking through walls, then the following is not so difficult to hypothesize. There is a higher dimension in which God knows everything that will happen.

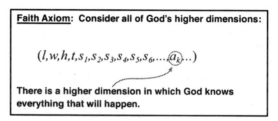

Faith Axiom: Consider all of God's higher dimensions:

$$(l, w, h, t, s_1, s_2, s_3, s_4, s_5, s_6, \ldots a_k \ldots)$$

There is a higher dimension in which God knows everything that will happen.

Figure 9.12 Foreknowledge in a higher dimension

Question 2: Prayer—Does God Answer Prayer? Can He Do So in Ways That Alter the Course of the Life of Mankind?

It has long been a debate among Christians and non-Christians whether God can answer prayer. In Chapter 6, we examined scientific evidence that God does indeed answer prayer. Sometimes we get a quick answer, sometimes we have to wait, and sometimes we get the answer "no." The answer is in God's will. We should always pray for an answer *in His will*. There is a higher dimension in which God can listen to prayer and alter the course of mankind, if it is in His will. Call this another faith axiom. The higher dimensions sum up all the theological and scientific information on the power of prayer. See also Psalm 16:11, Psalm 73:23-24, and 1 John 5:13-15.

$$(l, w, h, t, s_1, s_2, s_3, s_4, s_5, s_6, \ldots, a_k, a_{k+1}, \ldots)$$

Faith Axiom. There is a higher dimension in which God can listen to prayer, and alter the course of mankind, if it is in His will.

Figure 9.13 There is a higher dimension in which God can listen to prayer and alter the course of humankind, if it is in His will.

How can we think of or visualize how God might listen to prayer and thereby allow man to alter the course of humankind? The following diagram shows several alternate paths through each of several changes in life. Based on prayer, God allows different paths from one point to the next, even though He has foreknowledge of the effect of the overall path. God has built a plan for history that allows for a certain amount of *free play*. That free play can be altered in response to our prayers.

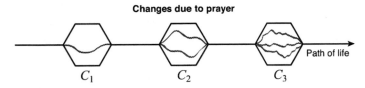

Changes due to prayer

Path of life

C_1 C_2 C_3

Figure 9.14 Different paths through God's plan for history, altered through prayer

Question 3: Pain and Suffering—Can God Redeem Pain and Suffering?

The answers we offer using higher dimensions may seem simplistic. But many difficult questions can be answered by thinking of God working in a higher dimension that we cannot understand in our present four-dimensional existence. Such an example is the problem of pain and suffering. Many books and articles are devoted to this issue in Christian thought. Whatever the result of our reasoning and seeking, the issue boils down to the one thing God asks of us: faith and trust. Although space limitations and the focus of this book prevent me from dealing with pain and suffering in depth, I am led to another faith axiom: *God can use adverse situations of pain and suffering to create good and to make us grow.* There is a higher dimension in which God can use pain and suffering to create goodness and redeem the paradox of the pain.

Faith Axiom. God can use adverse situations of pain and evil to create good and to make us grow. There is a higher dimension in which God can use pain and evil to create goodness and resolve the paradox of the pain.

$$(l, w, h, t, s_1, s_2, s_3, s_4, s_5, s_6, \dots, a_k, a_{k+1}, a_{k+2}, \dots, a_p, \dots)$$

Figure 9.15 A higher dimension exists in which God can create goodness from pain and suffering.

The following Bible verse supports this conceptualization of pain and suffering:

> **Rom 8:28:** *...God causes everything to work together for the good of those who love God and are called according to his purpose for them.*

See also Job 2:10 and Isaiah 55:8-9.

Question 4: Creativity—Can We Be Part of the Causality of Change of the Path of Mankind through the Creativity God Empowers in Our Intellect?

Closely related to God allowing prayer to alter the plan of history is the idea that God not only grants us free will, but also gives us the freedom to use His gift of creativity to alter the course of humankind. But God operates in a dimension where He knows what will happen. The situation is similar to the free play God grants us in prayer.

Faith Axiom. There is a higher dimension in which God allows man to use creativity to alter the path of life.

$$(l, w, h, t, s_1, s_2, s_3, s_4, s_5, s_6, \ldots, a_k, \ldots)$$

God allows humankind to use the creativity He has given us to alter the path of life.

Figure 9.16 There is a higher dimension in which God allows us to use creativity to alter the path of life.

Wormholes

We begin now to develop a series of ideas that allows me to create a theory about the end of the world as we know it, the end times as depicted in Revelation 21. This theory truly integrates mathematics, science, and religion. The first idea is that of a *wormhole*. Scientists sometimes consider higher dimensions to be modeled by parallel planes, called *parallel universes*. Passage from one dimension to another is virtually impossible. But scientists theorize that there may exist passages from one dimension to another. These passages are called *wormholes*.

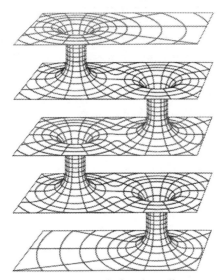

Figure 9.17 Tunnel-shaped wormholes that allow passage from one dimension to another

The planes represent higher and higher dimensions. The tunnel-shaped channels represent wormholes. It is theorized that passage between dimensions might be possible if an extreme amount of energy were available; humankind is not presently able to create that energy. But God has the energy to allow passage from one dimension to another via a wormhole! This idea will come to bear as our narrative progresses to Revelation 21 and the end times.

The Big Crunch and the New Creation

Cosmologists study the physical universe by examining history, structure, and constituent dynamics. Just as the BB was the beginning, scientists believe the *Big Crunch* (BC) to be the end of time as we know it. At the BB, the universe rushed into existence much like the rush of air into a balloon. Since then, the universe expanded much like the balloon continuing to be blown up. But physics indicates that the process will eventually come to a halt, and the universe will compress back onto itself, perhaps like the balloon bursting. Kaku says,

> For the cosmologist, the only certainty is that the universe will one day die. Some believe that the ultimate death of the universe will come in the form of the Big Crunch. Gravitation will reverse the cosmic expansion generated by the Big Bang and pull the stars and galaxies back, once again into a primordial mass. As the stars contract, temperatures will rise dramatically until all matter and energy in the universe are concentrated into a colossal fireball that will destroy the universe. All life forms will be crushed beyond recognition. There will be no escape....

The following is a brief sketch of Revelation 21. Think in the early stages of the end times where God takes all His followers away from this world with Him. That occurrence is called the *rapture*. Of those who remain, many eventually become followers of Christ, but only while they endure great tribulations. A great deal of time passes, Christ appears once again, and eventually the new followers of Christ are allowed to "escape" into a new world, or New Creation. That New Creation is a rectangular-shaped solid in which there is no night, and the physics of this present world are changed. It is, in essence, a new dimension. The theories of cosmology today suggest clues that a new world can exist as well as the ability to get there.

My theory is that God does have a plan to reverse the cosmic expansion—it is evidenced by Revelation 21. The notion of energy was not known in Bible times. The closest the Bible comes to discussing energy is through lightning (Matthew 24:27), thunder (Jeremiah 10:13), wind (Psalm 135:7), and fire (in the following passages). Although it is unlikely a fire could create enough energy for the BC, we can't deny that some unknown event could cause enough force to literally tear a hole in the dimension we inhabit today. But can we imagine God in a higher dimension having the heat and energy to cause the universe to contract and enact the BC? Time as we know it, created by God at the BB, may very well end at the BC. That is, time as God created it will be a closed interval.

We interpret or speculate that the subsequent passages offer strong Biblical support for the heat and energy needed for the BC to occur, but we do so by interpreting "fire" to represent that heat and energy:

2 Peter 3:5-7: *They deliberately forget that God made the heavens by the word of his command.... And God has also commanded that the heavens and the earth will be consumed by fire [heat and energy] on the day of judgment, when ungodly people will perish.*

Rev 16:8: *Then the fourth angel poured out his bowl on the sun, causing it to scorch everyone with its fire [heat and energy].*

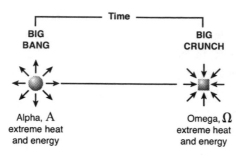

Figure 9.18 The time interval from the beginning (Big Bang) to the end (Big Crunch)

God's Followers at the Big Crunch

What happens to God's followers at the time of the BC? The late Gerald Feinberg, a physicist at Columbia University, said, "there is one, and only one, hope of avoiding the final calamity." He speculated that intelligent life, eventually mastering the mysteries of higher-dimensional space over billions of years, will use the other dimensions as an escape hatch from the Big Crunch.

God could be the designer of such a plan for His followers. We considered the possibility of wormholes between higher dimensions. String theorists hypothesize that if there were enough heat and energy, we could pass from one dimension to another through a wormhole. Who can provide that extreme energy? God, of course. Who will be allowed to escape through the wormholes to the next dimension? Why can't God allow them to be His followers? What will be the nature of this next dimension?

The Pieces of a Puzzle: String Theory, Higher Dimensions, and Revelation

Revelation, the last book in the New Testament, was written by John, thought by many to be one of Christ's disciples. John was exiled and placed in prison on the Greek island of Patmos. Through a vision, the resurrected Christ reveals to John a dream of hope: Christ will return (the second coming) and carry out the ultimate triumph of God over evil, the end of the earth as we know it, and the creation of the New Jerusalem.

Some Biblical scholars consider Revelation in a strict literal sense, whereas others view it as a metaphor filled with symbolism. Many believe it was written for Christians in ancient Rome; others believe fiercely that it gives us actual details that will happen within our life-times. Either way, the story of Revelation is difficult to interpret and understand. What follows is my interpretation of Revelation 21 using string theory, higher dimensions, and the Bible as pieces of a puzzle. In much the same way that I used mathematics to model growth of the evangelized population, I want to reinforce to you that this is *a model or an interpretation*. I am proposing a theory, a speculation or conjecture, in which I make no claim to absolute knowledge of what will happen. Put on your Technicolor imagination hat, and let's go for a ride in this theory.

Figure 9.19 Revelation, mathematics, higher dimensions, and string theory come together (drawing by David A. Hardy)

We will use higher dimensions to model the path of God's followers to the New Jerusalem described in Revelation. Let's look at some Biblical quotes. Early in the book, John quotes the word of Christ:

> **Rev 3:10-12:** *"Because you have obeyed my command to persevere, I will protect you from the great time of testing that will come upon the whole world to test those who belong to this world. I am coming soon. Hold on to what you have, so that no one will take away your crown. All who are victorious will become pillars in the Temple of my God, and they will never have to leave it. And I will write on them the name of my God, and they will be citizens in the city of my God—the new Jerusalem that comes down from heaven from my God. And I will also write on them my new name."*

The part of Revelation prior to Revelation 21 concerns warnings to seven churches that Jesus will return to vindicate the righteous and judge the wicked, followed by a portrayal of the rise of evil and then a vision of hope. Revelation 21 begins with a description of the New Jerusalem. That is where we start our interpretation. To aid your imagination, look at the previous figure and the one that follows this discussion as you read. The next quotes are by the author of Revelation, assumed to be John:

> **Rev 21:1-2:** *Then I saw a new heaven and a new earth, for the old heaven and the old earth had disappeared. And the sea was also gone. And I saw the holy city, the new Jerusalem, coming down from God out of heaven like a beautiful bride prepared for her husband.*

The rectangular solid in the previous figure can be thought of as the New Jerusalem. The verses imply that instead of God's followers going up to heaven, God is coming down to be with His followers. There is a parallelism here with Christ coming to earth to be with man.

> **Rev 21:3-4:** *I heard a loud shout from the throne, saying, "Look, the home of God is now among his people! He will live with them, and they will be his people. God himself will be with them. He will remove all of their sorrows, and there will be no more death or sorrow or crying or pain. For the old world and its evils are gone forever."*

This passage describes some aspects of the New Creation to add to those we considered before—only these are straight from the Bible. We begin to expect that eternity with God will be beyond our imagination.

> **Rev 21:5-8:** *And the one sitting on the throne said, "Look, I am making all things new!" And then he said to me, "Write this down, for what I tell you is trustworthy and true." And he also said, "It is finished! I am the Alpha and the Omega- the Beginning and the End. To all who are thirsty I will give the springs of the water of life without charge! All who are victorious will inherit all these blessings, and I will be their God, and they will be my children. But cowards who turn away from me, and unbelievers, and the corrupt, and murderers, and the immoral, and those who practice witchcraft, and idol worshipers, and all liars-their doom is in the lake that burns with fire and sulfur."*

Christ describes himself here as the alpha (A, first letter of the Greek alphabet; the beginning) and the omega (Ω, the last letter of the Greek alphabet; the end). Such an analogy is consistent with the fact that the New Testament was written in Greek. What follows are some of the rewards of heaven, "springs of the water of life," and inheriting "all these blessings." But a possible description of Hell also follows for those whose lives are full of sin; "their doom is in the lake that burns with fire and sulfur." This represents the burning earth.

Rev 21:10-14: *So he took me in spirit to a great, high mountain, and he showed me the holy city, Jerusalem, descending out of heaven from God. It was filled with the glory of God and sparkled like a precious gem, crystal clear like jasper. Its walls were broad and high, with twelve gates guarded by twelve angels. And the names of the twelve tribes of Israel were written on the gates. There were three gates on each side-east, north, south, and west. The wall of the city had twelve foundation stones, and on them were written the names of the twelve apostles of the Lamb.*

The angel who talked to me held in his hand a gold measuring stick to measure the city, its gates, and its wall. When he measured it, he found it was a square, as wide as it was long. In fact, it was in the form of a cube, for its length and width and height were each 1,400 miles. Then he measured the walls and found them to be 216 feet thick. [The angel used a standard human measure.]

These passages offer a physical description of the New Jerusalem, or heaven. Note that the shape is a rectangular solid, and it includes length, width, and height as in the earlier figure. The reference to "twelve" can be thought of as corresponding to the 12 tribes of Israel from the Old Testament. The dimensions of the rectangular solid are multiples of 12.

Rev 21:23-27: *And the city has no need of sun or moon, for the glory of God illuminates the city, and the Lamb is its light.... Nothing evil will be allowed to enter...*

This statement suggests that a new physics will prevail. Ross supports such a situation in the New Jerusalem, devoting an entire chapter on "Extra-Dimensionality and the New Creation!" He comments in reference to the cubical shape of the New Jerusalem that "If the familiar forces of gravity were in operation, this cube would collapse into a spherical shape, for in our universe, any material object with dimensions exceeding 300 miles across would be pulled by gravity into a more or less spherical shape."

> **Rev 22:12-13:** *"See, I am coming soon, and my reward is with me, to repay all according to their deeds. I am the Alpha and the Omega, the First and the Last, the Beginning and the End."*

Revelation 22 relates Christ's promise to return in the second coming. If the Omega is the Big Crunch, and to create the BC God provides the extreme heat and energy, then God may create a wormhole for His followers to pass to the New Jerusalem, and the heat and energy will consume the earth as we know it now.

Let's assemble the puzzle. God brings heat and energy at the end times. God creates a wormhole, which is a passage from earth's dimension as we know it to a higher dimension, the New Jerusalem. String theorists hypothesize that with an extreme amount of energy, people can pass from one dimension to another. God is omnipotent; He has the energy. God then allows His followers to pass through the wormhole to the New Jerusalem, which is the rectangular solid in the previous and following figures. Because of the energy that has been enacted by God, the six string dimensions are activated from their present inactive state. In what follows, a new physics prevails, the six string dimensions stay active, and there is no more night. The light of the New Jerusalem is Christ, and it shines forever.

The previous figure shows the Earth before the BC, with the wormhole. The next figure shows the Earth being consumed by the heat and energy near the time of the BC. God's followers have passed through the wormhole.

Figure 9.20 Heat and energy destroying Earth near the time of the Big Crunch (drawing by David A. Hardy)

We might even construe these events to be God's grand resolution of a kind of paradox between the Big Bang and the Big Crunch.

Figure 9.21 A paradox of the Big Bang and the Big Crunch, resolved by the New Jerusalem

In summary, we have used the concepts of higher dimensions as metaphor to examine concepts such as the soul, heaven, and the existence of God. We also use the metaphor to examine the Trinity, as well as how the fourth dimension might be interpreted in terms of the fiery furnace, the transfiguration, and Christ's appearance to His disciples after His resurrection. We also brought the concept of higher dimensions to bear on the notions of the foreknowledge, prayer, pain and suffering, and creativity. Finally, we integrated ideas from mathematics—namely, higher dimensions—with those of science, relativity, quantum physics, string theory, and the philosophy of paradox, ending of this chapter with a new theory of the Christian view of the end times as depicted in Revelation 21.

Conclusion

On some of these points, you may question whether using higher dimensions as a metaphor isn't a cop-out, or a quick way to resolve some very thought-provoking issues in the Christian faith. To me, they are not cop-outs, for two reasons. First, as a person who tries to grapple with issues of the faith by a dynamic resonance of belief and questioning, I come closer to belief through the insight of this thinking. Second, by discovering that strings—rapid vibrations between opposing entities analogous to paradox—are at the foundation of our biological being, and that at the apex of our being is our psycho-spiritual growth that comes from a different kind of thinking about paradox, I come to an amazing acceptance that the higher dimensions are more than metaphor: They permeate our existence. They also provide for me a certain peace regarding the acceptance of the fact that God has an infinite mystery about Him, and we will never come to grips with all of it in this life. But I can't wait to see that infinite mystery unfold in heaven. I'm reminded of a Biblical passage:

> **1 Cor 13:12:** *Now we see things imperfectly as in a cloudy mirror, but then we will see everything with perfect clarity. All that I know now is partial and incomplete, but then I will know everything completely, just as God now knows me completely.*

Chapter 10

Numerical Applications of Mathematics in the Bible

But by far the most memorable teacher of my undergraduate days, and the one who had the most impact on my career, was the well-known number theorist and haberdasher, Professor Natalie Attired. I vividly recall her greatest lecture—a very clear and satisfying proof of the Chinese Remainder Theorem. I must admit, however, that an hour after she proved it, I was hungry for more mathematical knowledge.

—**William Dunham**
(Professor of Mathematics,
Muhlenberg College, Allentown, PA)

Yes, mathematicians can be humorous; and fun is the tack we take in this chapter, especially after all the math, physics, and Biblical dialogue. This chapter doesn't cover curve fitting, statistics, or string theory. It doesn't even discuss any meaningful Christian apologetics. It's for the mathematics fan (purist) who enjoys finding math in all kinds of Biblical and theological places.

Numerics: The Number 7

As you read the Bible, you'll begin to notice that starting with Genesis in the Old Testament, the occurrence of *seven* or *seventh* jumps out at you. As a math fan, seeing those 7s was a message that got my attention.

If you're curious about the number of occurrences of the number 7 in the Bible, you can use a concordance to look up *seven*, *sevenfold*, *sevens*, *seventh*, *seventeen*, *seventeenth*, and *seventy* and see for yourself how often the number flows through the Bible from Genesis through Revelation. The Old Testament was written in Hebrew. In those times, the number 7 meant perfection. Thus, it seems reasonable that it found its way into the Bible.

There are 66 books in the Bible; *seven* or *seventh* occurs in the first 21 out of 39 books of the Old Testament (also note that $21 = 7 \cdot 3$). The first book it does not occur in is "Song of Songs," or "Song of Solomon." In the New Testament, *seven* or *seventh* occurs in the first five books and in three others, the last being Revelation:

> First use in the Bible: **Gen 2:2:** *"On the 7th day, having finished his task, God rested from all his work."* This is the second chapter in the Bible. God completed creation and rested.

> Last use in the Bible: **Rev 21:9:** *"Then one of the 7 angels who held the 7 bowls containing the 7 last plagues came...."* This quote is in the next-to-last chapter in the Bible, revealing the reuniting of the Church with Christ. Of interest is that the first occurrence of a seven is in the second chapter from the front of the Bible and the last occurrence is the second chapter from the back.

The number **7** is used in almost *600* passages in the Bible. Revelation includes 40 uses of *seven* and 5 uses of *seventh*. Some scholars assert that the occurrences of *seven* in the Bible implies perfection, fullness, and completeness.

Here are two more uses of 7 as multiples:

Parables: Jesus spoke or conveyed 70 parables:
$70 = 7 \cdot 10$

Recorded Miracles: Jesus performed 35 recorded
miracles: $35 = 7 \cdot 5$

You will find frequent occurrences of other numbers in the Bible such as 8, 12, and 40, although we will not consider them in detail here. For example, the number for God's people is 12: There were 12 tribes in Israel, and 12 apostles started the church.

Other Numerical Applications

All kinds of interesting numerical observations are hidden in the Bible. Here is a helpful memory device: The phrases *OLD TESTA-MENT* and *NEW TESTAMENT* each have two words containing 3 and 9 letters. When we write 3 and 9 together, we get 39, the number of books in the Old Testament. When we multiply $3 \cdot 9 = 27$, we get the number of books in the New Testament. But perhaps most notable is the occurrence of the numbers 666, 777, and 888.

When you mention the number 666, you conjure controversy. The book of Revelation introduces the number 666 in the following scripture:

> **Rev 13:18:** *Wisdom is needed to understand this. Let the one who has understanding solve the number of the beast, for it is the number of a man. His number is* **666**.

In some ancient manuscripts, 666 is given as 616. Mankind has spent the last 2,000 years trying to attach meaning to this scripture and the number, often referred to as the "mark of the beast." It is my opinion that the understanding of this will have to wait until we pass into

God's dimension or during the end times and that no worthwhile interpretation exists. Nevertheless, if you do some browsing on the Internet, you will quickly become embroiled in the controversy.

We consider a few possibilities here, but only because they are interesting mathematically:

- In the Greek alphabet, each letter is assigned a specific number. Thus, a correspondence or function is defined. The letter *stigma* is no longer part of the Greek alphabet, but when it was, it corresponded to the number 6. If we read 666 as "six, six, six" in Greek, we can read the number in ancient Greek manuscripts as "stigma, stigma, stigma." Such a word in Greek would mean "the imposter of Christ."
- Some consider Nero, emperor of the Roman Empire (AD 54–68), to be the most evil man who ever lived, next to Hitler. He is credited with the murder of his mother and one of his wives as well as the slaughter of Christians for entertainment. It is asserted that the Hebrew letters of Nero's name represent numbers that total 666. This can be done by taking the numerical values of the words *Nero Caesar* written in Hebrew. This doesn't seem as logical to me because it takes two words to get the conversion; and according to Martin Livio, "the spelling in Hebrew of the word Caesar actually omits a letter, of value 10, from the more common spelling."

The following are some other interpretations of 666 as related to 777 and 888, but they are somewhat speculative.

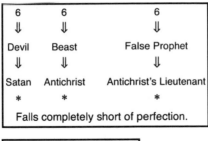

Figure 10.1 Symbolic representations of 666, 777, and 888

There have also been a few humorous attempts at interpreting 666. The following are purely tongue-in-cheek, as created by Lawrence Braden, a student at St. Paul's School, Concord, NH:

Notation for 666 Interpretation

DCLXVI	Roman numeral of the Beast
666.0000	Number of the high-precision Beast
0.666	Number of the milli-Beast
1010011010	Binary of the Beast
666 + 666i	Complex number of the Beast
1-666	Area code of the Beast
00666	Zip code of the Beast
1-900-666-6666	Live Beast! One-on-one pacts! Only $6.66
$665.95	Retail price of the Beast
666k	Retirement plan of the Beast
6.66%	Beastly interest rate

Sorry, but I could not resist adding a few of my own, as follows:

Notation for 666 Interpretation

Phillipi 666	Gasoline of the Beast
Appian 666	Way of the Beast
666	Direct Satellite channel of the Beast
\$666,000,000,000	The Beastly price of selling your soul
$\sqrt{443,556}$	The Beast in square root clothing is still the Beast
$6^{10^{100}}$	The Googol Beast
666	The Beastly total of three games of bowling, as coveted by your author
–0.0176416458	The "Sine" of the Beast
666π	The Beast's favorite dessert
$\sum_{i=1}^{\infty} 666i$	The sum of all of Beastly desires
66 Z 666	License number of the Beast in the state of Indiana

Medians in the Bible

The *median* of a set of numbers is the middle number in the set, if the number of elements in the set is odd. In such a case, the middle number is an *element* of the set. For example, the median of the set

{98, 99, 870, 91, 106, 98, 90}

is 98. It helps to first write the numbers in order:

{90, 91, 98, 98, 99, 106, 870}

When the number of elements in the set is even, the *median* of the set is the average of the two middle numbers. In such cases, the median may or may not be in the set. For example, the median of the set {69, 80, 61, 63, 62, 65} is 64:

$$\left(\frac{63+65}{2} = 64 \right)$$

It turns out that Psalm 118 is the middle or median chapter in the Bible. There are 594 chapters in the Bible before Psalm 118 and 594 chapters after Psalm 118. Interestingly, Psalm 117 is the shortest chapter in the Bible, and Psalm 119 is the longest. What is the median of the Bible in terms of verses? It is Psalm 118:8: "It is better to trust the Lord than to put confidence in people."

Some translations of the Bible are missing a verse: John 5:4. In the New Living Translation, the writing skips from the third to the fifth verse. But in the New American Standard Bible that verse is

> **John 5:4:** ...*for an angel of the Lord went down at certain seasons into the pool, and stirred up the water; whoever then first, after the stirring up of the water, stepped in was made well from whatever disease with which he was afflicted.*

Were that verse included, the preceding numerics would change— the Bible would have no middle verse. But either way, the facts would still be interesting Biblically and mathematically.

The Golden Ratio or Section

Studied by the Greeks before the time of Euclid, the number Φ, described with the Greek letter *phi*, is called the *golden ratio* or *golden section* and is defined as follows:

$$\Phi = \frac{1 + \sqrt{5}}{2} \approx 1.618$$

That is, Φ = the golden ratio = the golden section.

As far back as 2,000 years ago, artists, sculptors, and architects regarded the way to divide a line that is "most pleasing to the eye" to be such that the ratio of the long side to the short is Φ. Let's derive Φ according to this description. Consider the following line.

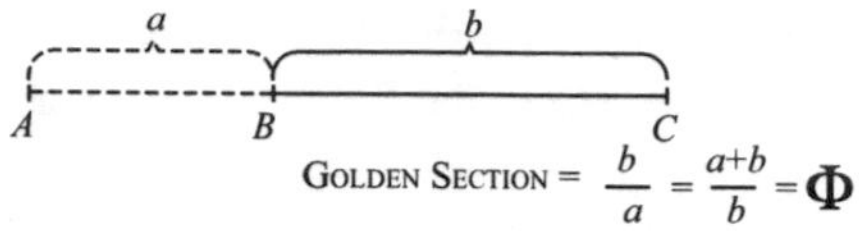

Figure 10.2 Visualizing the golden ratio, Φ

It is cut in such a way that the ratio of the length b of segment BC to the length a of segment AB is the same as the ratio of the entire segment AC, $a + b$, to b. Then, setting $b / a = x$ in the following equation, we get

$$\frac{b}{a} = \frac{a+b}{b}$$

$$\frac{b}{a} = \frac{a}{b} + 1$$

$$x = \frac{1}{x} + 1 \qquad \text{Substituting } x \text{ for } \frac{b}{a}$$

$$x^2 = 1 + x$$

$$x^2 - x - 1 = 0$$

Solving this equation using the quadratic formula, we have

$$x = \frac{1 \pm \sqrt{5}}{2}$$

The only positive root is the number $\dfrac{b}{a}$, or Φ.[*] Thus,

$$\Phi = \frac{1 + \sqrt{5}}{2} \approx 1.618$$

---

[*]In some discussions, Φ is defined as

$$\Phi = \frac{-1 + \sqrt{5}}{2} \approx 0.618$$

Note that Φ is close to the ratio of 5 to 3, or 1.6, and

$$\frac{1}{\Phi} = \Phi - 1 \approx 0.618$$

That is, the reciprocal of the golden ratio can be found by subtracting 1 from the number. Also, $\Phi^2 \approx 2.618$. That is, the square of the golden ratio Φ is found by adding 1 to the number Φ.

Mario Livio has written a lively, readable, and comprehensive book called *The Golden Ratio*. No one knows for sure the derivation of the word *golden* to describe this number, but it may have originated in the second edition of a book by Martin Ohm published in 1835 entitled *Die Reine Elementari-Mathematik* (*The Pure Elementary Mathematics*). It arises in all kinds of applications, not just in mathematics and physics. According to Livio, "biologists, artists, musicians, historians, architects, psychologists, and even mystics have pondered and debated the basis of its ubiquity and appeal." We will consider a few of these applications here and one Biblical application.

The golden ratio has had extensive historical interest to Greek mathematicians and others. It was felt to possess aesthetic and artistic properties. Rectangles for which the ratio of the length to the width is Φ were considered to be "the most pleasing to the eye" and called "golden rectangles."

Following are drawings of several rectangles together with the ratios of length to width. Which rectangle is most pleasing to your eye? Also, in the figure, you see a drawing of the Greek Parthenon with an imposed rectangle that is considered golden—although, according to Livio, it is difficult to say for sure that the Parthenon architects used Φ in its design (the theory is not particularly well supported by the Parthenon's actual dimensions).

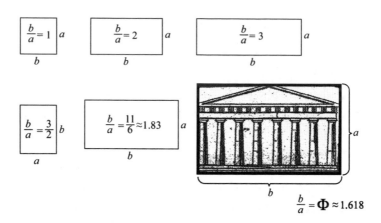

Figure 10.3 The golden ratio in architecture

We can find almost any number in an architectural design by juggling figures. Pick your number, and take some measurements. Then, do some calculations, but choose only those results close to the number, and write a treatise on the "divinity" of your results. This can be referred to as *juggling* or *stretching* the data. For example, pick the number 2.7. Suppose we took two measurements in some piece of architecture, divided them, and got an answer that was approximately 2.7. To announce to the world that "God divined 2.7 in this piece of architecture," in the absence of other information, would be a stretch of the data.

The Greek mathematician Pythagoras started a brotherhood called the Pythagoreans, which was a school of historians of mathematics (*c.* 550–300 BC). The following figure shows the symbol of the Pythagoreans: the five-pointed star pentagram, which is derived from the regular pentagon shown on the right.

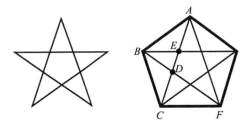

Figure 10.4 The golden ratio in a regular pentagram

The golden ratio Φ can be found in the pentagon as follows:

$$\Phi = \frac{AC}{AB} = \frac{AC}{AD} = \frac{AD}{AE} = \frac{AE}{DE}, \quad \text{and so on.}$$

Indications of the golden ratio have shown up in the work of artists such as Georges Seurat, Albrecht Durer, and Leonardo da Vinci, as well as in the architecture of the Great Pyramid in Egypt. Index cards of dimensions 3 x 5 and 5 x 8 are such that the ratio of length to width is close to Φ. But although that fact is interesting, those cards simply represent a way of visualizing Φ. I doubt that the inventor of that size of the cards had Φ in mind.

One Biblical application of the golden ratio relates to the dimensions of the Ark of the Covenant, although such a result might be considered a stretch. In Exodus 25:10, God mandates to Moses to "Make an ark of acacia wood—a sacred chest $3\frac{1}{4}$ feet long, $2\frac{1}{4}$ feet wide, and $2\frac{1}{4}$ feet high." The ark was a rectangular box that would hold the stones inscribed with the Ten Commandments.

Figure 10.5 The golden ratio in the dimensions of the Ark of the Covenant

Let's examine the ratio of the width to length:

$$\frac{3\frac{1}{4}}{2\frac{1}{4}} = \frac{3.25}{2.25} \approx 1.444$$

This ratio is somewhat close to the golden section:

$$\Phi = \frac{1+\sqrt{5}}{2}$$

But only God knows for sure, because mathematics in Biblical times had not developed irrational numbers like $\sqrt{5}$. Even I would accept the criticism of number juggling or stretching regarding this "discovery."

Hidden Codes in the Bible

In the interest of providing information, I consider two controversial efforts regarding the possibility of the existence of hidden codes in the Bible. At the outset, I must assert that the word of God would not have survived for thousands of years if it was as elusive as the Bible codes discussed here. But from the point of view, "Is it mathematically interesting or informative?" we take a brief look.

Several books have been written on the subject over the years, including at least two in the past decade. Ultimately, no one knows if these theories have any validity; but from a purely mathematical perspective, they're interesting to consider as we look at the broader way that numbers interact with our world.

The Work of Ivan Panin

The first assertion of hidden codes in the Bible comes from the work of Ivan Panin. Born in Russia in 1855, Panin emigrated to the United States, entered Harvard University in 1878, and graduated with a bachelor's degree in 1882. His mathematical abilities are a source of controversy: Some claim he was brilliant, whereas others claimed he struggled for mediocre grades at Harvard. Panin was first known for a letter written to the *New York Sun* entitled *The Inspiration of the Scrip-*

tures Scientifically Demonstrated, which was published in the issue of Sunday, 19 November 1899; he published several other similar books and articles.

Gematria is numerology of the Greek and Hebrew alphabets; its proponents use it to predict human relationships or future events. Gematria was sometimes combined with astrology to form an ancient branch of superstitious learning. Looking for hidden codes is an aspect of gematria. The study of these codes can involve the assignment of number values to letters of the alphabet, thus giving a word a numerical value by finding the sum of the number values of its letters. Panin did gematrial studies of the Bible, mainly considering or looking for multiples of 7. In the Hebrew alphabet, each letter is assigned a specific number, although a few letters are assigned two numbers. In the Greek alphabet, each letter is assigned to exactly one number.

We know historically that most of the Old Testament was written in Hebrew and most of the New Testament in Greek. Thus, one way to perform a gematrial study on a passage is to consider the numerical values as they were assigned historically to letters in both alphabets. For example, let's see how we can use the Greek alphabet to find a word's value. The following table shows the number assignments historically used for the Greek alphabet. We consider these because the Greek letters are used so much in mathematics and because to each letter is assigned exactly one number—that is, we have a function.

Letter	Greek Symbol: uc, lc	Numerical Value
Alpha	A, α	1
Beta	B, β	2
Gamma	Γ, γ	3
Delta	Δ, δ	4
Epsilon	E, ε	5
Stigma*	No longer used	6
Zeta	Z, ζ	7
Eta	H, η	8
Theta	Θ, θ	9
Iota	I, ι	10

Letter	Greek Symbol: uc, lc	Numerical Value
Kappa	K, κ	20
Lambda	Λ, λ	30
Mu	M, μ	40
Nu	N, ν	50
Xi	Ξ, ξ	60
Omicron	O, o	70
Pi	Π, π	80
Koppa*	No longer used	90
Rho	P, ρ	100
Sigma	Σ, σ	200
Tau	T, τ	300
Upsilon	Y, υ	400
Phi	Φ, φ	500
Chi	X, χ	600
Psi	Ψ, φ	700
Omega	Ω, ω	800
Sampsi*	No longer used	900

*Around 6 B.C., these letters were removed from the Greek alphabet.

For example, let's find the numerical value of the word *Jesus* (meaning *Savior*), or Ιησουσ in Greek:

I	η	σ	o	υ	σ
⇓	⇓	⇓	⇓	⇓	⇓
10 +	8 +	200 +	70 +	400 +	200 = 888

Figure 10.6 The numerical value of Jesus

Using the table, we assign the numerical value to each letter, and then we add the numbers to get the word's number value: 888.

Many critics of Panin's work and other code studies dispute whether the inspiration of the Holy Bible by God can be conclusively proven by such studies. Inasmuch as it is beyond our capability to interview the original authors of the Bible, much less to scientifically verify any kind of communication between them and God, then no, it cannot be *proven* by gematria that the Bible is divinely inspired. At the *most*, gematria may show that the Bible is a unique book or at least interesting purely from a mathematical standpoint.

Panin spent most of his life attempting to show by gematria that God divinely inspired the Bible. On the surface, the patterns he uncovered in the Greek New Testament (mainly involving the number 7) seem incredible if they are not inspected further. In the New Testament passage Matthew 1:1-17, we find the following counts of things that are multiples of seven:

Words: $7^2 = 49$
Words beginning with a vowel: $7 \cdot 4 = 28$
Words beginning with a consonant: $7 \cdot 3 = 21$
Letters in the entire vocabulary: $7 \cdot 38 = 266$
Letters: $7 \cdot 20 = 140$
Consonants: $7 \cdot 18 = 126$
Words that occur more than once: $7 \cdot 5 = 35$
Words occurring only once: $7 \cdot 2 = 14$

Perhaps the most notable of Panin's critics in modern times is a professor at the Australian National University named Brendon McKay. McKay points out that in the Greek text of Mark 16:9-20, Panin found more than 150 "features" involving the number 7, similar to those just shown for the passage in Matthew. This is but a small fraction of the total number of features of *any* number—not just seven—that could be found in a Scripture passage. We could focus on any one number and try to find it in the number of letters in the vocabulary, the number of letters in the passage, the number of words that begin or end with a consonant, the number of words that begin or end with a vowel, the number of proper names, and so on. Some of these features would be multiples of the number we chose, simply by chance. We could go through the entire Bible (or any other writing), noting every minute numerical feature for every letter and every word, compile the results in a huge table, and then focus on a few numerical features that seem to be connected and claim that this was the result of divine intervention. The situation is similar to the number-juggling we discussed with the golden section, Φ.

There are many variants of the Greek text of the New Testament; they generally differ in small details. When we find numerical patterns in letters and words, variant readings that do not change the meaning of the passage can easily change the numerical patterns. For example, one of the patterns Panin found was a multiple of seven in the number of words in the vocabulary of a certain passage. The addition or deletion of just one small word in the vocabulary destroys the pattern. Panin apparently made the *a priori* assumption that the numerical patterns in the Bible were divinely ordained, and he used variant Greek New Testament versions to buttress his assumption. He believed that the combination of variants that manifested the patterns he sought was the unique Word of God, and he published a *Numeric Greek New Testament* based primarily on Brooke Westcott and Fenton Hort's *The New Testament in the Original Greek* but freely borrowing from other versions, which fit his assumptions to the letter.

On his web page, McKay writes of the findings of Ken Smith of Brisbane who, "…did an investigation which proves our point forcefully. Panin's report on the last twelve verses of Mark begins with the observation that there are $175 = 25 \times 7$ words in the Greek text." Smith collected a large number of editions and counted the words in that passage. His findings follow. From the discrepancy in the word count, we might suspect that many other things are wrong with Panin's analysis.

Edition	Date	Words
Elzevir's edition of Textus Receptus	1624	166
Wilson	1864	165
Alford	1874	166
Westcott and Hort	1881	172
Weymouth	1886	167
Nestle	1898	168
Souter for Accepted Version	1902	168
Souter for Revised Version	1902	168
Nestle	1904	168
Souter	1910	168
Huck	1936	167

Edition	Date	Words
Souter	1947	169
British and Foreign Bible Society	1958	168
Tasker	1961	165
Nestle/Aland	1975	170
Huck/Greeven	1981	168

None of these editions has the right number of words for Panin's claims. What chance do they have for Panin's assertions concerning letter counts or numerical values? We conclude that Panin designed the patterns he found. Thus, although many of Panin's "discoveries" are interesting, I'm led to the conclusion that some stretching took place with his work, and I certainly attach no divine insight to his results.

Equidistant Letter Sequences (ELS)

The second assertion of hidden codes in the Bible began with the work of Elijah ben Solomon (1720–97, a Jewish scholar, a.k.a. the Gaon of Vilna), who asserted that "all that has happened or will happen is in the Torah," the first five books of the Bible. Some believe he was saying that codes in the Bible reveal the secrets of the universe. Later, Sir Isaac Newton (1642–1727) tried relentlessly to find hidden codes in the Bible, but to no avail.

In 1997, a book by Michael Drosnin, *The Bible Code*, sparked a renewed interest in the subject. It appeared on the *New York Times* bestseller list for several months and led to a flurry of other books about secret codes in the Bible—and tremendous controversy.

The basis of finding these codes is the search for equidistant letter sequences (ELS). Here is a brief explanation of the ELS process according to Rick Milne:

A computer finds the first letter of a word, and then begins counting until it finds the next letter of the word. This becomes a "skip code." Then, using that skip code, it counts to see if the third letter of the word is found at the

> same interval. So it would start by skipping every other letter, then every two letters, then every three letters until it finds a "skip" that spells out the word.

Then, the skip code is used to find other meaningful words within a related rectangular array of letters.

The *Notices of the AMS* (American Mathematical Society) is a journal held in high regard by mathematicians. In the September 1997 issue, Allyn Jackson reviewed Drosnin's book, stating that "The sad fact is that this book is a series of wild, unfounded claims based on stretching statistical evidence to the breaking point. ...one cannot avoid the conclusion that he [Drosnin] lacks the mathematical and statistical background that would bring some depth.... He is deluded by his ignorance."

It is true that Drosnin used his Bible code to predict the assassination of Israeli leader Yitzak Rabin, the name of his assassin, and the year he was killed, but the code offered no precise details about where or when the assassination would occur. Another mathematician, Doron Witztum, used the code to predict that Winston Churchill would be assassinated, but he died of natural causes. If we predicted that a U.S. president would be assassinated, the prediction would have about a 20 percent probability of being correct. Compared to the more overt prophecies of the Bible, the Bible code falls way short in making predictions.

Drosnin's work is a whirlpool of controversy. On the positive side, recent use of his code in *The Bible Code Digest* to analyze Isaiah 53 reveals many theological wordings—more than we would find in a non-encoded, normal piece of literature by a comparison of 50.0 percent to 19.4 percent—and many of these make reference to *Jesus*. Isaiah is an Old Testament book, so these wordings might be considered prophetical. Some of them are *Jesus is truth, True messiah, It is not an echo, but the Father in them is Living Water, operating powerfully,* and *I will remember the resurrection of Jesus, who is alive.*

Is the preceding explanation sufficient to convince you of the elusiveness and uncertainty of this procedure? McKay, Milne, and Ralph Muncastor refute the notion of secret Bible codes. To me, this concept is not worth the space to fully develop in depth, nor is it worth your time. (Again, were the word of God that elusive, it would not have survived to this day!) You may find such a quest fun or interesting, but I consider it much like doing sudoku or crossword puzzles.

As to the best-selling book *The Da Vinci Code*, by Dan Brown, keep in mind that it is fiction, so any information you think might be factual must be verified elsewhere. The author makes many references to math terms like the golden ratio and the Fibonacci sequence, but those reveal no valid or insightful Biblical information.

Exercises

Up to this point in the book, it has included no exercises, although you might construe some topics to be research questions. If you enjoy math, you may be missing some exercises to work out on your own, so I've provided some here. Notice the Bible translation being used, and remember that modern translations often convert ancient measures to American measures like feet and yards.

1. The Size of Noah's Ark

In Biblical measures, it is thought that 1 cubit ≈ 18 in, roughly the distance from the elbow to the tips of the fingers. (There is also a "long" cubit, which is about 21 in.) The dimensions of Noah's Ark are given in Genesis 6:15 (KJV) as follows: "The length of the ark shall be three hundred cubits, the breadth [width] of it fifty cubits, and the height of it thirty cubits."

   a. What were the dimensions of Noah's ark in inches? in feet?

   b. Find the ratio of the breadth to the height. What might God think was most pleasing to the eye about the ark?

2. Goliath's Height

In Biblical measures, it is thought that 1 span = $\frac{1}{2}$ cubit. The giant Goliath's height according to 1 Samuel 17:4 (KJV) "was six cubits and a span."

a. What was the height of Goliath in inches?

b. What was the height of Goliath in feet?

3. The Remnant

Revelation 14:1 says, "Then I saw the Lamb standing on Mount Zion, and with him were 144,000 who had his name and his Father's name written on their foreheads." The 144,000 represent the so-called "remnant" of believers at the end times who are believers in Christ, have survived the persecutions on earth, and will reap eternal benefits.

a. Express 144,000 in terms of a multiple of 12.

b. If each tribe of Israel has the same number of people, how many are in each tribe? See if this is to happen by checking Revelation 7:4-8.

4. The New Jerusalem

Revelation 21:15-17 says, "The angel who talked to me held in his hand a gold measuring stick to measure the city, its gates, and its wall. When he measured it, he found it was a square, as wide as it was long. In fact, it was in the form of a cube, for its length and width and height were each 1400 miles. Then he measured the walls and found them to be 216 feet thick (the angel used a standard human measure)." In the end times, the 144,000 will live in a new city called The New Jerusalem.

a. Find the volume of the New Jerusalem. How much volume in cubic feet will each of the 144,000 people have if there are no other residents?

b. Estimate the volume of a normal-sized 6-foot person. How many of these "normal-sized" people will fit in New Jerusalem, if they are crammed together as close as possible?

c. What might the results of questions a and b tell us about the number of people at the end times?

d. In the New International Version (NIV) of the Bible, the length, width, and height of the New Jerusalem are described as being "1200 stadia," a unit used in Bible times. What is 1 stadia in miles? kilometers?

5. Solomon's Temple

1 Kings 6:16-20 says, "He partitioned off an inner sanctuary—the 'Most Holy Place' at the far end of the Temple. It was 30 feet deep and was paneled with cedar from floor to ceiling. The main room of the Temple, outside the Most Holy Place, was 60 feet long. Cedar paneling completely covered the stone walls throughout the Temple, and the paneling was decorated with carvings of gourds and open flowers.

Solomon prepared the inner sanctuary in the rear of the Temple, where the Ark of the Lord's covenant would be placed. This inner sanctuary was 30 feet long, 30 feet wide, and 30 feet high. Solomon overlaid its walls and ceiling with pure gold."

The dimensions and shape of the Most Holy Place are in the same proportion as The New Jerusalem.

a. Find the volume of the Most Holy Place.

b. Describe that proportionality.

6. The Pillars of Solomon's Temple

According to Jeremiah 52:20-21, "The bronze from the two pillars, the water carts, and the Sea with the twelve bulls beneath it was too great to be measured. The things had been made for the Lord's Temple in the days of King Solomon. Each of the pillars was 27 feet tall and 18 feet in circumference. They were hollow, with walls 3 inches thick."

a. Calculate the volume of the empty space inside each pillar.

b. Calculate the volume of the bronze in each pillar.

The volume V of a cylinder with height h whose base is a circle with radius r is given by $V = \pi r^2 h$.

7. Imperfection

a. Read Revelation 11:3 (NLT). The 1,260 days referred to are how many years?

b. How is $3\frac{1}{2}$ related to 7, the perfect number in Biblical terms?

The number $3\frac{1}{2}$ is associated Biblically with imperfection, incompletion, and evil. The period written about by John in Revelation 11:3 is a time of many unfavorable situations. See if you can describe them by reading the following references.

c. Revelation 11:2

d. Daniel 12:7 (In this passage, this unfavorable period is predicted.)

e. Revelation 13:5

Answers

1. a. Length: 5400 in. or 450 ft, width: 900 in. or 75 ft; height: 540 in. or 45 ft
   b. Breadth/height = 900/540 or about 1.667. The ratio approximates the golden ratio, so God might have thought this ratio to be "most pleasing to the eye."

2. 117 in. or 9.75 ft

3. a. $144,000 = 12 \cdot 12,000 = 12^2 \cdot 1000$
   b. From Revelation 7:4-8, we know there are 12 tribes of Israel, so the number of people in each tribe would be the answer (a) divided by 12, or 12,000.

4. a. Total volume: $(1400 \cdot 5280 - 2 \cdot 216)^2 (1400 \cdot 5280) \approx 4.0 \times 1020$ ft3 :volume per person: approximately 2.8×10^{15} ft^3
   b. Approximately 10 ft^3; 4.0×10^{19} people
   c. There could be up to about 4.0×10^{19} people
   d. One stadia is about 1 1/6 mi, or about 1.9 km

5. a. 27,000 sq ft
   b. A cube: length/width = length/height = width/height = 1

6. 580 ft^3

7. a. 3 1/2 yr
   b. Half of 7, half of perfection
   c. Occupation/destruction of Jerusalem
   d. Shattering of the "holy people"; rule by "the beast"

Chapter 11

Gödel's Hurdle and the Issue of Proof

I confess that Fermat's Theorem as an isolated proposition has very little interest for me, because I could easily lay down a multitude of such propositions, which one could neither prove nor dispose of.

—Karl Friedrich Gauss

Over the mathematical ages, many problems or statements in mathematics have remained unproven. It turns out that one of the most astounding mathematical results of the twentieth century shows that there exist statements that are either true or false, but *cannot* be proven one way or the other from the axioms of the system. Such a mathematical frustration—the fact that it's possible for something to be true but remain unproven—serves as a metaphor for the final faith axiom of this book. As mathematicians, we want to prove everything we can using the axioms available to us, but we have to live with the reality that there are statements we can't prove as theorems. As human beings seeking answers in many areas of our lives, we know how comfortable or uncomfortable we may be making a leap of faith based on evidence. But, can we prove "everything"? And if not, how do we come to peace with what we do know?

Famous Unsolved Problems

To prepare for a metaphor between a famous mathematical result and its connection to faith, let's consider some famous problems from mathematics: some unsolved, and some solved only recently after years of effort.

Goldbach's Conjecture

A classic problem, unsolved as of this writing, is *Goldbach's Conjecture*. Examine the following, and look for a pattern:

$6 = 3 + 3$
$8 = 5 + 3$
$10 = 5 + 5$
$12 = 7 + 5$
$14 = 7 + 7$
$16 = 11 + 5$

...

Note that each number on the left is even and can be expressed as a sum of two odd prime numbers. (A *prime* number is a positive integer greater than 1 that has exactly two divisors, itself and 1. The numbers 2, 3, 5, 7, 11, 13, and 17 are examples of prime numbers.) The following is Goldbach's Conjecture:

> *Goldbach's Conjecture.* Every even integer greater than 4 can be expressed as a sum of two odd prime numbers.

This problem is easy to convey but has remained unsolved for many years. It has been verified for even numbers up to $4 \cdot 10^{14}$. Do you want to earn an honorary doctorate really fast? Then come up with a proof.

Riemann Hypothesis

In contrast to Goldbach's Conjecture, the Riemann Hypothesis is difficult to express, although it also involves prime numbers.

In June 2004, Louis De Branges of Purdue University claimed to have proved the Riemann Hypothesis, and as of this writing, his proof is being critiqued by other mathematicians. De Branges has announced a proof a number of other times, but all of his previous proofs were found faulty by reviewing mathematicians. (You can check the proof here: http://www.math.purdue.edu/~branges/riemannzeta.pdf.)   The proof is 41 pages in length. In De Branges' defense, he is well-known for having proved another famous and at-the-time unsolved problem, the *Bieberbach Conjecture*. That proof was also preceded by many failed attempts. Such are the ways of advanced mathematics.

Famous Solved Problems

Now, let's consider some famous problems that have been solved recently.

The Four-Color Problem

If you examine a colored map, you will notice that typically, each country or state is a different color than its immediate neighbor, although countries or states whose borders meet at a point may have the same color. The four-color problem arose in 1852, when Francis Guthrie wrote to his brother that four colors were always sufficient. Try this for yourself. See if you can use four colors for each of the following maps under the criteria given. In fact, both can be colored with four colors.

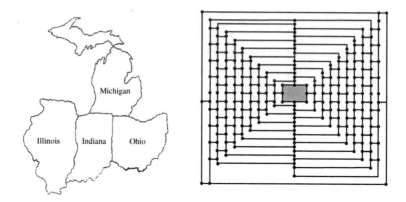

Figure 11.1 Coloring a map using only four colors

Well-known mathematician Arthur Cayley presented the problem to the London Mathematical Society in 1878:

> Any map in a plane can be colored using four colors in such a way that regions that have a common boundary, other than a single point, will not have a common color.

The problem remained unproven until Kenneth Appel and Wolfgang Haken developed a proof in 1976. The proof filled an entire book, published in 1989, and required an extensive amount of computer time.

Fermat's Last Theorem

Most geometry students are familiar with the Pythagorean Theorem. It states that the square of the hypotenuse (the side opposite the right angle) is equal to the sum of the squares of the other two sides.

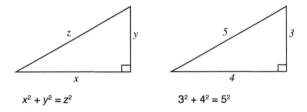

$x^2 + y^2 = z^2$ $3^2 + 4^2 = 5^2$

Figure 11.2 The Pythagorean Theorem

For example, in a 3-4-5 right triangle, $3^2 + 4^2 = 5^2$. If we were to ask ourselves if there are any integer solutions of the equation $x^2 + y^2 = z^2$ that are nontrivial (consisting of numbers other than combinations of 0s and 1s), the 3-4-5 case ($x = 3$, $y = 4$, and $z = 5$) gives us an answer. There are infinitely many solutions, among them 9-40-41, 8-15-17, and 5-12-13.

Suppose we raise the exponents to 3 and ask if there are any non-trivial integer solutions of the equation $x^3 + y^3 = z^3$. The only solutions that come to mind are the trivial cases $1^3 + 0^3 = 1^3$, $0^3 + 1^3 = 1^3$, and $0^3 + 0^3 = 0^3$.

In fact, there are no nontrivial solutions. If we raise the exponents to any natural number $n > 2$, the situation of no nontrivial solution still prevails. This is known as *Fermat's Last Theorem*:

The equation $x^n + y^n = z^n$ has no nonzero integer solutions for x, y, and z, when n is an integer such that $n > 2$.

Pierre de Fermat (1601–1665) was a French lawyer who considered himself to be an amateur mathematician. His "recreational dabbling" in mathematics resulted in two important branches of mathematics: one known as number theory and the other probability theory.

Number theory arose from the many unsuccessful attempts to prove Fermat's Last Theorem. Fermat wrote a note in the margin of a book in 1630 to the effect that he had discovered a truly remarkable proof, but it was too lengthy to place there. Such an act can drive mathematicians to great efforts, as evidenced by the growth of number theory over the past 400 years. I wonder how much energy the mathematical world would have generated if he had also written in the margin, "I have discovered a proof of the existence of God, but it is too lengthy to place here!"

British mathematician Andrew Wiles, while working at Princeton University, announced that he had discovered a proof. With the help of other mathematicians such as Gerd Faltings and Richard Taylor, the proof was simplified and accepted in mathematical circles by 1995. The result made front pages around the world and inspired a television documentary and a bestselling book.

Related to this act of Fermat is a humorous story about well-known British mathematician G. H. Hardy (1877–1947), told by George Polya with religious overtones:

Hardy, an atheist, stayed in Denmark with Bohr (another mathematician) until the very end of the summer vacation, and when he was obliged to return to England to start his lectures there was only a very small boat available (there was no airplane traffic at the time). The North Sea can be pretty rough and the probability that such a small boat would sink was not exactly zero. Still, Hardy took the boat, but sent a postcard to Bohr: "I proved the Riemann hypothesis. G. H. Hardy." You re not laughing? It is because you don't see the underlying theory. If the boat sinks and Hardy drowns, everybody must believe that he has proved the Riemann hypothesis. Yet God would not let Hardy have such a great honor and so he will not let the boat sink.

Gödel's Hurdle

Kurt Gödel (1906–1978; last name rhymes with *hurdle*) is considered to be the greatest logician of the twentieth century, or indeed any century. It was not so much that he proved a previously unproven result, it was that he proved a fact that was totally unanticipated and virtually unbelievable: his *Incompleteness Theorem*. This result rocked the worlds of mathematics and philosophy.

I once took an advanced mathematical logic course in graduate school where we spent an entire semester on the proof of Gödel's Incompleteness Theorem. When finished, I felt as though I had traversed "Gödel's hurdle." Here we will briefly consider the meaning of the theorem, but we definitely will not consider its proof. Then, we will draw a metaphor to the Christian faith.

A mathematical system involves assumptions, called *axioms*, made about the set of objects that is the framework of the system. A *statement* made within the structure of the mathematical system is either true or false; it cannot be both. Using rules of logic and reasoning, statements are then proved using the axioms and previously proven statements. The proven statements as well as the axioms are referred to as *theorems*. We can visualize theorems as follows.

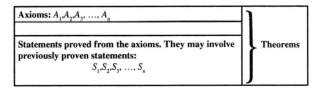

Figure 11.3 Theorems from axioms

A desired property of a mathematical system is that it be *consistent*, meaning there is no statement Q such that both Q and (*not* Q) are theorems of the mathematical system. Note how this takes us again close to the notion of paradox. We can visualize consistency as follows. In a consistent mathematical system, the following *cannot* happen.

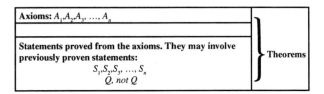

Figure 11.4 Not possible in a consistent mathematical system

For example, suppose a mathematical system is such that both $2 + 3 = 5$ and $2 + 3 \neq 5$ are theorems. Then we know the system is not consistent.

Another desirable attribute of a mathematical system is that it be *complete*. A mathematical system is complete if for every statement S describable in the system, either S is a theorem, or (*not S*) is a theorem.

We can visualize completeness as follows. Let S be a statement describable in the system. That means it is either true or false but not both. Then, either

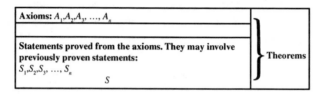

Figure 11.5 S is a theorem.

or

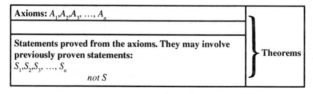

Figure 11.6 (*not S*) is a theorem.

Gödel proved two theorems, but we shall consider only the first. In both theorems, his proofs pertain to a mathematical system rich enough in structure to at least involve arithmetic: the natural numbers

$$\{1, 2, 3, 4, 5, \ldots\}$$

together with their operations, such as addition, subtraction, multiplication, and division. The axioms allow the existence of a number like 1 and then a different number $1 + 1$, which we call 2. Then, another number exists, $2 + 1$, which is defined as 3, and so on. The process continues and produces all the natural numbers.

The following is Gödel's First Incompleteness Theorem:

In a consistent mathematical system M rich enough to
at least involve the numbers of arithmetic, there exists
a sentence S such that S is true, but S cannot be proven
within the system. That is, M is incomplete.

Saying this another way, you can have a statement S where either
it is true or its negation (*not S*) is true, but you can't prove either one
within the system. Thus, the system is incomplete. Suppose such a sentence turned out to be Goldbach's Conjecture, G. We know that either
G is true or its negation (*not G*) is true. But we can never prove either
statement.

Gödel released this result and its proof in 1931. The proof is loosely
based on the Barber's Paradox we discussed in Chapter 2, but in no
way is Gödel's Incompleteness Theorem a paradox or a contradiction.
It is authentic and valid mathematics, which humbled the mathematical
community at the time.

In their book *Gödel's Proof*, Ernest Nagel and James R. Newman
stated that

A climate of opinion was thus generated [in mathematical
circles] in which it was tacitly assumed that each sector of
mathematical thought can be supplied with a set of axioms
sufficient for developing systematically the endless totality
of true propositions [another word for statements] about
the given area of inquiry.

Gödel's paper showed that this assumption is untenable.
He presented mathematicians with the astounding and
melancholy conclusion that the axiomatic method has
certain inherent limitations, which rule out the possibility
that even the ordinary arithmetic of the integers can ever
be fully axiomatized. What is more, he proved that it is
impossible to establish the internal consistency of a very
large class of deductive systems—elementary arithmetic
for example—unless one adopts principles of reasoning

so complex that their internal logical consistency is as
open to doubt as that of the systems themselves. In the
light of these conclusions, no final systemization of many
important areas of mathematics is attainable, and no
absolutely impeccable guarantee can be given that many
significant branches of mathematical thought are entirely
free from internal contradiction.

You might ask, "If we find such a statement S, that is unprovable, why don't we just add it to the system M and move on?" Nagel and Newman add, "Even if the axioms of arithmetic are augmented by an indefinite number of other ones, there will always be further mathematical truths that are not formally derivable from the augmented set." This is an utter frustration to the mathematical notion of assuming as little as possible in an axiom set and then proving its theorems.

What Is the Metaphor to the Christian Faith?

The metaphor of Gödel's theorem leads us to a final faith axiom: No matter what we assume in our human knowledge, we can never assume enough to prove it all. We can't prove it all in mathematics. We can't prove it all about God. A certain amount of mystery will always remain. All our axioms about God and His ways are never enough to prove or reveal all His knowledge. James Nickel adds,

Man's reason is not autonomous; man is not omniscient
[he does not have total knowledge].... To the Biblical
Christian, Gödel's results confirm the truth that scripture
had always proclaimed. Autonomy belongs to the Biblical
God alone. Whenever man tries to construct any system
of thought without reference to this God, it will ultimately
fall short.

Gödel even felt his result carried with it religious significance. Gödel was by nature a reclusive man. He wasn't inclined toward intellectual competitions with his fellow mathematicians, although he cherished a deep friendship with Einstein. According to Rebecca Goldstein in her book *Incompleteness: The Proof and Paradox of Kurt Gödel,*

The most important moments of his career were, in fact, those about which we know nothing: the moments of intuitions or thought-experiments or God-knows-what that brought him to the proof itself. The overall strategy of the proof is astoundingly simple, the details that had to be worked out are astoundingly complicated, and both astounding features make us wish we knew more about how he came up with it all.

I for one would have been fascinated to be inside Gödel's head to connect with the creative processes behind this amazing result. Maybe I can do that in Heaven as well as enjoy many other higher-dimension revelations.

Bibliography

Abbott, Edwin. *Flatland: a Romance of Many Dimensions.* Amherst, NY: Prometheus Books, 2005.

Alexander, Denis. *Rebuilding the Matrix: Science and Faith in the 21st Century.* Grand Rapids, MI: Zondervan, 2001.

Appel, Kenneth, and Wolfgang Haken. "Every Planar Map Is Four Colorable." *Contemporary Mathematics* 98 (1989).

Aristotle. *Nicomachean Ethics.*

Avi-Yonah, Michael. *The Jews of Palestine: A Political History from the Bar Kokhba War to the Arab Conquest.* Oxford: Blackwell, 1976.

Barrett, David B., George T. Kurian, and Todd M. Johnson. *World Christian Encyclopedia: A Comparative Survey of Churches and Religions in the Modern World.* 2nd ed. Oxford; New York: Oxford University Press, 2001.

Benson, Herbert, Jeffery A. Dusek, Jane B. Sherwood, Peter Lam, Charles F. Bethea, William Carpenter, Sidney Levitsky, et al. "Study of the Therapeutic Effects of Intercessory Prayer (STEP) in Cardiac Bypass Patients." *American Heart Journal* 151 (2006): 934-942.

Bittinger, Marvin. *One Man's Journey Through Mathematics.* Boston: Addison-Wesley, 2004.

Bittinger, Marvin L., David Ellenbogen, Judith Beecher, and Judith Penna. *Algebra and Trigonometry: Graphs and Models.* 3rd ed. Boston: Pearson Addison Wesley, 2006.

Buchanan, Mark. "The Benefit of the Doubt." *Christianity Today* (April 3, 2000).

Buechner, Frederick. *Wishful Thinking: A Seeker's ABC.* Rev. and exp. ed. San Francisco: HarperSanFrancisco, 1993.

Byrd, Randolph C. "Positive Therapeutic Effects of Intercessory Prayer in a Coronary Care Unit Population." *Southern Medical Journal* 81, no. 7 (1988): 826-9.

Campbell, Joseph. *The Power of Myth.* New York, NY: Anchor Books, 1991.

Cha, K. Y., and Daniel Wirth. "Does Prayer Influence the Success of In Vitro Fertilization-embryo Transfer? Report of a Masked Randomized Trial." *The Journal of Reproductive Medicine* 46 (2001): 781-787.

Chartrand, Gray, Albert D. Polimeni, and Ping Zhang. *Mathematical Proofs: A Transition to Advanced Mathematics.* Boston: Pearson Education, 2003.

Collins, Francis S. *The Language of God : A Scientist Presents Evidence for Belief.* New York: Free Press, 2006.

Craig, William Lane. "On Being a Christian Academic." Paper given at the National Christian Faculty Leadership Conference, Washington, D.C., June 2004.

————. *Reasonable Faith: Christian Truth and Apologetics.* Rev. ed. Wheaton, IL: Crossway Books, 1994.

Curry, Dayna, and Heather Mercer with Stacy Mattingly. *Prisoners of Hope: The Story of Our Captivity and Freedom in Afghanistan.* New York: Doubleday; Colorado Springs, CO: WaterBrook Press, 2002.

Dossey, Larry. *Healing Words: The Power of Prayer and the Practice of Medicine.* New York, NY: Harper Mass Market Paperbacks, 1997.

Drosnin, Michael. *The Bible Code.* New York: Simon & Schuster, 1997.

Edwards, Jonathan. *The Works of Jonathan Edwards.* New York: Garland Pub. Co., 1987.

Ehrman, Bart D. *Misquoting Jesus: The Story Behind Who Changed the Bible and Why.* New York: HarperSanFrancisco, 2005.

Eldredge, John. *Waking the Dead.* Nashville: Thomas Nelson Publishers, 2003.

Furlow, Leslie, and Josie Lu O'Quinn. "Does Prayer Really Help?" *Journal of Christian Nursing* 19, no. 2 (2002): 31-34.

Geisler, Norman L., and Paul K. Hoffman, eds. *Why I Am a Christian: Leading Thinkers Explain Why They Believe.* Rev. and exp. ed. Grand Rapids, MI: Baker Books, 2006.

Gershman, Jacob. "Sound Science?" *The New York Sun,* June 9, 2004, sec. 1.

Goldstein, Rebecca. *Incompleteness: The Proof and Paradox of Kurt Gödel.* New York: W. W. Norton, 2005.

Goswami, Bijoya, with David K. Wolpert. *The Human Fabric: Unleashing the Power of Core Energy in Everyone.* Austin, TX: Aviri, 2004.

Greene, Brian. *The Elegant Universe: Superstrings, Hidden Dimensions, and the Quest for the Ultimate Theory.* New York: W. W. Norton, 1999.

————. *The Fabric of the Cosmos.* New York: Alfred A. Knopf, 2004.

Hansen, Richard P. "Biblical Paradox Offers an Alternative to 'How-To' Sermons." *Christianity Today International/Leadership Journal* XXI, no. 1 (2000): 54.

Harris, Paul "Exposed: conman's role in prayer-power IVF 'miracle'" *The Observer,* Guardian Newspapers Unlimited, May 30, 2004.

Harris, William S., Manohar Gowda, Jerry W. Kolb, et al. "A Randomized, Controlled Trial of the Effects of Remote, Intercessory Prayer on Outcomes in Patients Admitted to the Coronary Care Unit." *Archives of Internal Medicine* 159 (1999): 2273-2278.

Hodge, David R. "A Systematic Review of the Empirical Literature on Intercessory Prayer." *Research on Social Work Practice* 17 (2007): 174-187.

Holy Bible: New Living Translation. Wheaton, IL: Tyndale House Publishers, 1st ed., 1996; 2nd ed., 2004.

Kaku, Michio. *Hyperspace: A Scientific Odyssey through Parallel Universes, Time Warps, and the Tenth Dimension.* New York: Anchor Books, 1995.

Katz, Victor J. *A History of Mathematics.* 3rd ed. Boston: Pearson Addison-Wesley, 2008.

Krucoff, Mitchell W., Suzanne W. Crater; et al. "Music, Imagery, Touch, and Prayer as Adjuncts to Interventional Cardiac Care: The Monitoring and Actualisation of Noetic Trainings (MANTRA) II Randomized Study." *The Lancet* 366 (2005): 211-217. See also www.thelancet.com.

Leibovici, L. "Effects of Remote Intercessory Prayer on Outcomes in Patients with Bloodstream Infection." *British Medical Journal* 323 (2001): 1450-1451.

Lewis, C. S. "The Efficacy of Prayer." *Fern-seed and Elephants.* UK: Fount, 1975.

———. *God in the Dock.* Grand Rapids, MI: William Eerdmans Pub. Co., 2000.

———. *The Great Divorce: A Dream.* San Francisco: HarperSanFrancisco, 2001.

———. *Of This and Other Worlds.* Edited by Walter Hooper. London: Collins, 1982.

———. *The Weight of Glory and Other Addresses.* San Francisco: HarperSan-Francisco, 2001.

Lindsell, Harold. *The Battle for the Bible.* Grand Rapids, MI: Zondervan Pub. House, 1976.

Livio, Mario. *The Golden Ratio: The Story of Phi, the World's Most Astonishing Number.* New York: Broadway Books, 2002.

McKay, Brendan. "Ivan Pavin [*sic*] and the Gospel of Mark." http://cs.anu.edu.au/~bdm/dilugim/panin_mark.html.

McLaren, Brian. *The Last Word and the Word After That.* San Francisco: Jossey-Bass, 2005.

———. *The Story We Find Ourselves In: Further Adventures of a New Kind of Christian.* San Francisco: Jossey-Bass, 2003.

McVey, Steve. *Grace Rules.* Eugene, OR: Harvest House, 1998.

Metzger, Bruce Manning. *The Text of the New Testament: Its Transmission, Corruption, and Restoration.* 4th ed. New York: Oxford University Press, 2005.

Moise, Edwin. *Elementary Geometry From an Advanced Standpoint.* Reading, MA: Addison-Wesley Pub. Co., 1962.

Moreland, J. P. *Love Your God With All Your Mind.* Colorado Springs, CO: Navpress, 1997.

Morris, Henry M. *Many Infallible Proofs: Practical and Useful Evidences of Christianity.* San Diego: Creation-Life Publishers, 1974.

Nagel, Ernest, and James R. Newman. *Gödel's Proof.* New York: New York University Press, 1958.

Neuhouser, David L. *Open to Reason.* Upland, IN: Taylor University Press, 2001.

Nickel, James D. *Mathematics: Is God Silent?* Vallecito, CA: Ross House Books, 2001.

O'Laire, S. "An Experimental Study of the Effects of Intercessory Prayer on Self-Esteem, Anxiety, and Depression." *Alternative Therapies* 3, no. 6 (1997): 38-53.

Paley, William. *Natural Theology: Or, Evidences of the Existence and Attributes of the Deity.* Edinburgh, UK: Lackington, Allen and Co., and Jamie Sawers, 1818.

Panin, Ivan. "The Inspiration of the Scriptures Scientifically Demonstrated." Letter to the editor. *The New York Sun*, November 19, 1989.

Parker, David C. *The Living Text of the Gospels*. Cambridge, UK; New York: Cambridge University Press, 1997.

Peck, M. Scott. *Golf and the Spirit*. New York: Three Rivers Press, 1999.

Peterson, Eugene. *The Message*. Colorado Springs, CO: Navpress, 2002.

Phillips, Tony. "The Great Moon Hoax." Science@NASA, 2001. http://science.nasa.gov/headlines/y2001/ast23feb_2.htm.

Polya, George. "Some Mathematicians I Have Known." *The American Mathematical Monthly* 76, no. 7 (1969): 752-753.

Ross, Hugh. *The Creator and the Cosmos: How the Greatest Scientific Discoveries of the Century Reveal God*. 3rd exp. ed. Colorado Springs, CO: NavPress, 2001.

Rucker, Rudy. *The Fourth Dimension: A Guided Tour of the Higher Universes*. Reprint ed. Houghton Mifflin, 1985.

Russell, Bertrand. *On God and Religion*. Edited by Al Seckel. Buffalo, NY: Prometheus Books, 1986.

Sicher, F., E. Targ, D. Moore, and H. S. Smith. "A Randomized Double-Blind Study of the Effects of Distant Healing in a Population with Advanced AIDS: Report from a Small Study." *The Western Journal of Medicine* 53 (1998): 116-121.

Stewart, Ian. *Flatterland: Like Flatland Only More So*. Cambridge, MA: Perseus Pub., 2002.

Stott, John, and Mark Noll. *Your Mind Matters: The Place of the Mind in Christian Life*. Leicester, England: InterVarsity Press, 2007.

Strobel, Lee. *The Case for Christ: A Journalist's Personal Investigation of the Evidence for Jesus*. Grand Rapids, MI: Zondervan, 1998.

Swindoll, Charles R. *The Grace Awakening*. Nashville: W Pub. Group, 2003.

Talbert, Richard J. A., in collaboration with Roger S. Bagnall et al, eds. *Barrington Atlas of the Greek and Roman World*. Princeton, NJ: Princeton University Press, 2000.

Walker, S. R., J. S. Tonigan, W. R. Miller, S. Comer, and L. Kahlich. "Intercessory Prayer in the Treatment of Alcohol Abuse and Dependence: A Pilot Investigation. *Alternative Therapies* 3, no. 6 (1998): 79-86.

Westcott, B. F., and F. J. A. Hort. *The New Testament in the Original Greek*. New York: Harper & Brothers, 1881.

Willard, Dallas. *Hearing God: Developing a Conversational Relationship with God*. Downers Grove, IL: InterVarsity Press, 1999.

———. *Hearing God Through the Year*. Compiled and edited by Jan Johnson. Downers Grove, IL: InterVarsity Press, 2004.

Williams, Daniel Day. *Interpreting Theology, 1918-1952*. London: SCM Press, 1953.

Yancey, Philip. *Reaching for the Invisible God: What Can We Expect to Find?* Grand Rapids, MI: Zondervan Publishing House, 2000.

Afterword

The thing is to understand myself, to see what God really wishes me to do; the thing is to find a truth which is true for me, to find the idea for which I can live and die.

—Søren Kierkegaard

To those who are followers of Christ, I hope you found ways to strengthen your faith by providing a fresh look using mathematics. To those who are not followers of Christ, it is my hope that the evidence provided by the mathematics may bring you closer to the joy of becoming a follower of Christ.

The following lists some of my important insights, or nuggets. Some of these insights are theological or philosophical, some are analytical, some are metaphorical; but all of them were gauged within a framework of mathematics. I state them from my heart in order to inform and inspire you to seek answers to your own questions by using the many references in this book, the Bible, prayer, and Christian literature and by finding a church that encourages open questioning. You can find God through these questions, especially when they lead to answers!

- Mathematicians make assumptions or axioms and then try to prove as many theorems as they can from the axioms using rules of logic. Faith axioms are assumptions that have intellectual consequences to the spiritual side of life. *Assuming the existence of God* is such a faith axiom. Pursue faith axioms in an atmosphere of questioning. Be a hunter, a detective, a stalker of evidence. Be passionate about finding truth. In this sense, be a true skeptic.

- Learning more about apologetics can answer many questions about Christianity and satisfy each person's need to use their brain to enhance their faith. In the process, you have more evidence to make a faith decision and are more prepared to share and defend your faith to others.

- Embracing paradox is a key to psycho-spiritual growth. Just as paradox is a key to psycho-spiritual growth, so is proof by contradiction a key method of proof in mathematics. Mathematicians would like to give up the use of proof by contradiction, but it is too powerful. Similarly, the embrace of paradox is too powerful to give up as we pursue psycho-spiritual growth.
- Christ is God's ultimate resolution of the paradox of humankind being caught between self-centeredness and God-centeredness.
- The integration of science (mathematics) and the Christian faith can reveal huge insights into the way God has made the world.
- Every Christian should attempt to integrate their faith with their profession. C. S. Lewis implores us to become "little Christs" in our endeavors.
- No matter what we assume in our human knowledge, we can never assume enough to prove it all. We can't prove it all in mathematics. We can't prove it all about God. A certain amount of mystery will always remain. All our axioms about God and His ways are never enough to prove or reveal all His knowledge. God's will is a not a puzzle to be solved, it is mystery to be savored.
- A study of the probability of prophecy reveals astounding evidence of the reliability of the Bible. No other book in history makes hundreds of predictions hundreds of years apart, none of which have yet failed to come true. The character of God is not simply the Bible; the Bible is the most extraordinary way to God's character.

If you have just completed this book but are a not a follower of Christ, it has been my joy to present this material—and, yes, I have an ulterior motive, which I suspect you have discovered by now. I'm going to cut to the chase; you can agree with me, you can not agree with me, but I am going to reveal my heart. I would love for you to become a follower of Christ, and I hope I have written a compelling enough case for you to at the very least investigate further. I hope you are nearing the conclusion that what you are missing that will make your heart right is becoming a follower of Jesus Christ. Brian McLaren leads me to the question, "Wouldn't you like to live with the wind at your back rather than in your face, and wouldn't you like to be on the side of God's ongoing creation rather than on the side of destructive human selfishness?" I want to invite you to join this incredible journey.

We have considered many apologetic arguments, most within a framework of mathematics. If you still have questions, I encourage you to encounter the many books referenced throughout this book, especially those by Yancey, Strobel, Boyd, Peck, Lewis, Neuhouser, and Craig—and, of course, the Holy Bible. There I would start with the Gospel of John, followed by Romans, Hebrews, Philippians, and 1, 2 Timothy. Along the way, try Proverbs—I love the one-liners. In addition, seek counsel with your pastor and/or a Christian friend, especially one who is well-studied in the faith.

If some aspect of this book has been stimulating to your heart, and you know as you are reading these words that something is moving you, then what follows is the next step I would recommend you take. It is something I have done. You may not be able to put two words together in a prayer, but here is a prayer that you could use:

God, I know I have been filled with self-centeredness, following my ways and not yours. I accept the amazing gift of grace your son, Christ, made for me when He died on the cross for my sins. I also rejoice in the hope given to me when He rose from this death. Now, please come and live within me and fill me with your power. I want your wind at my back. I want to be on the side of your ongoing creation rather than on the side of destructive human selfishness. Reveal to me your desire for my heart and fill me with the resolve to act in your behalf. I'm on fire to follow you!

The faith decision is profound! Please do not make the decision unless your *mind* is convinced intellectually that you truly want to follow Christ. Please do not make the decision unless it comes from within your *heart*. Please do not make the decision without a *will* to follow Christ, a bias to action. In the process, you are carrying out the faith equation:

Faith = (Mind) + (Heart) + (Will)

If you are already a follower of Christ, use the framework of mathematics to bring a fresh approach to your knowledge of apologetics. The goal was to strengthen your faith, bring you more courage and hope, and maybe stimulate you to look a little deeper into the foundation of what you believe so that ultimately you can affect other people. All Christians should endeavor to strengthen the intellectual aspects of the faith, the *mind* part of the faith equation. As John Stott says, "Knowledge is indispensable to Christian life and service. If we do not use the mind which God has given us, we condemn ourselves to spiritual superficiality."

I encourage all my readers to seek further sources such as a Bible-teaching church or friends who are open to questions. If you are a scientist (mathematician) of any type, consider researching and/or writing about integrating science (mathematics) and faith.

Let's consider one last paradox. On one side of the vibration, it amazes me that mathematicians can engage in the utter abstractness of the symbols, axioms, theorems, and statements in their fields but not attempt to connect them to the spiritual side of life. We draw mathematics from realities like the motion of automobiles and the flight of a projectile. Is it that much of a stretch to draw notions of God from the realms of nature, medicine, music, science, beauty of the arts, and the mind's creativity, and then pursue them intellectually?

Metaphysics studies the nature of reality, such as the relationship between mind and matter, substance and quality, fact and value. If you have ever studied Greek philosophy, you may have considered a type of metaphysics called *Platonism*. Developed by Plato, it asserts that ideas as they are known by the mind are actual realities. For example, to a Platonist, the color red is a concept of the mind that is a reality unto itself apart from the objects that might be used to build the concept such as red bricks, apples, crayons, stop signs, and clothes.

On the other side of the paradox, mathematicians tend to be Platonists. The realities they deal with on a daily basis are the myriad of symbols like π, ∞, $\sum$, $\frac{dy}{dx}$, and $\lim_{h \to 0}$, together with the mathematical systems that involve them, like The Real Number System, Infinite Dimensional Vector Spaces, General Topology, and Modern Abstract Algebra.

But, as Craig writes, "On Platonism, God is reduced to but one necessary being among many, an infinitesimal part of reality, most of which exists utterly independently of Him. Such a metaphysical pluralism seems incompatible with the Christian doctrine of God who alone exists necessarily and eternally and is the Creator of all reality outside Himself." Stated another way, mathematicians deal with so many abstract realities that they might throw the concept of God into all the those realities and downplay Him from His top rank as the creator of all reality outside Himself, including mathematics.

C. S. Lewis wrote, "I believe that any Christian who is qualified to write a good popular book on any science [or mathematics] may do much more by that than any directly apologetic work. What we want is not more little books about Christianity, but more little books by Christians on other subjects—with their Christianity *latent*." It is my prayer to have written "a little book on mathematics with my Christianity latent." Thank you for sharing my journey!

For more information on the book, check the website http://www.thefaithequation.com. It contains an ongoing list of questions for research and discussion, as well as appearances by your author, and more depth on the mathematics. I welcome interaction with my reader. You can reach me at faithequation@gmail.com.

Index

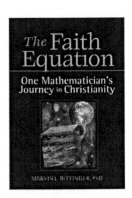

Printed in the United States
83500LV00001B/94-999/A

9 781933 669076